Wittgenstein's Lectures on the Foundations of Mathematics

Also available:

Wittgenstein's Lectures: Cambridge, 1930–1932
From the Notes of John King and Desmond Lee
Edited by Desmond Lee

Wittgenstein's Lectures: Cambridge, 1932–1935
From the Notes of Alice Ambrose and Margaret Macdonald
Edited by Alice Ambrose

Wittgenstein's Lectures on Philosophical Psychology 1946–47
Notes by P. T. Geach, K. J. Shah, and A. C. Jackson
Edited by P. T. Geach

WITTGENSTEIN'S LECTURES
on the Foundations of Mathematics
Cambridge, 1939

FROM THE NOTES OF

R. G. BOSANQUET, NORMAN MALCOLM, RUSH RHEES, and YORICK SMYTHIES

EDITED BY CORA DIAMOND

The University of Chicago Press
Chicago and London

The University of Chicago Press, Chicago 60637
The University of Chicago Press, Ltd., London
 Originally published 1976
University of Chicago Press edition 1989
Printed in the United States of America
98 97 96 95 94 93 92 91 90 89 5 4 3 2 1

Library of Congress Cataloging-in-Publication Data

Wittgenstein, L. (Ludwig), 1889–1951.
[Lectures on the foundations of mathematics, Cambridge, 1939]
Wittgenstein's lectures on the foundations of mathematics. Cambridge, 1939 : from the notes of R. G. Bosanquet, Norman Malcolm, Rush Rhees, and Yorick Smythies / edited by Cora Diamond.
p. cm.
Reprint. Originally published: Ithaca, N.Y. : Cornell University Press, 1976.
Includes bibliographical references.
ISBN 0-226-90426-1 (alk. paper)
1. Mathematics—Philosophy. I. Bosanquet, R. G. II. Diamond, Cora. III. Title.
[QA8.6.W57 1989]
510'.1—dc20 89-37788
CIP

♾ The paper used in this publication meets the minimum requirements of the American National Standard for Information Sciences—Permanence of Paper for Printed Library Materials, ANSI Z39.48-1984.

Contents

Editor's Preface

Wittgenstein wrote a great deal on the foundations of mathematics between 1929 and 1944. During this period, he discussed the philosophical problems of the foundations in several sets of lectures at Cambridge; among the last was that given in the Lent and Easter terms of 1939. Norman Malcolm has described these lectures in *Ludwig Wittgenstein: A Memoir;* there is another brief description, by D. A. T. Gasking and A. C. Jackson, in "Ludwig Wittgenstein." [1] The lectures, which were given twice a week, lasted two hours, and Wittgenstein spoke entirely without notes. The notes published here are based on those taken by students at the lectures.

Those present at the lectures included, besides Malcolm and Gasking, R. G. Bosanquet, J. N. Findlay, Casimir Lewy, Marya Lutman-Kokoszynska, Rush Rhees, Yorick Smythies, Stephen Toulmin, A. M. Turing, Alastair Watson, John Wisdom, and G. H. von Wright.[2] I had at my disposal the notes of Bosanquet, Malcolm, Rhees, and Smythies (which I refer to as B, M, R, and S). A pirated version of Malcolm's notes was published under the title *Math Notes* in San Francisco in 1954, and Bosanquet's version was in private circulation for a time.

The four manuscripts from which I worked were of different sorts and presented different problems. Bosanquet's version was the fullest, but in writing up his notes Bosanquet had edited them for his own purposes, rearranging material, filling in details, and altering grammar and style. Rhees and Malcolm had written up

1. *Australasian Journal of Philosophy,* 29 (1951).

2. There are also references in the lectures to Cunningham and Prince, whom I have not been able to identify.

their notes with a certain minimal degree of editing and interpretation; only in the case of Smythies was I working with notes in the form in which they were made during the lectures, entirely unedited and sometimes barely legible. None of the four versions included all thirty-one lectures.[3] In many passages three or four versions agreed quite closely; in others there were discrepancies, more or less considerable.

My aim in preparing the text was to produce from those four versions a single version which was both readable and as accurate as possible, given the difficulties; footnotes and variant readings were kept to a minimum. No single version was taken as *the* basic text. Rather, each passage is based on a comparison of all the available versions of that passage. Where two or more versions agreed in some point, I normally took them to be correct in that respect. As a consequence, there are sentences to which nothing in any of the four versions exactly corresponds. Where there was no agreement and it was necessary to decide on a single version, I tended, though not invariably, to follow Rhees or Smythies, since, in other contexts, each of them agreed with at least one other version more often than did Bosanquet or Malcolm; and Bosanquet's was the most highly edited version, Malcolm's often the briefest. Certain choices, especially those concerned with the order of the material, had to be made with no adequate basis in any version; the only 'method' here was that of determining what made the best sense and was at the same time consistent with the evidence. It will be clear, then, that the accuracy of the text varies and depends to a certain extent on the accuracy of my ear, and also that many passages could have been handled differently. I have indicated in the footnotes those passages in which there are special difficulties of some sort. The absence of a footnote does not imply that the text is based on clear and conclusive evidence, only that other ways of dealing with the material would not differ significantly. Some repetitive passages have been cut. The use of quotation marks follows the conventions in *Remarks on the Foundations of Mathematics*.

3. Lectures I–VIII are based on B, M, and S; IX–XVI on B, M, R, and S; XVII on B, R, and S; XVIII–XXV on B, M, R, and S; XXVI–XXVII on B, M, and R; XXVIII–XXIX on M and R; XXX–XXXI on M and S.

The accuracy of the text is important for two quite different reasons. Whoever had said these things, they would still illuminate the philosophical issues, and they would still have been spoken in a highly characteristic voice, in language whose forcefulness conveys the *kind* of thought that went into them. Even if the lectures were anonymous, then, there would be good reason to want the words accurately given, the voice not muffled, nor the language distorted. But we also want an accurate record of the lectures because they are Wittgenstein's, and so may cast light on other things he said. A great deal of caution must, however, be used before anything in the text here can be taken as 'giving Wittgenstein's views' or even as giving good evidence for some particular interpretation of what he says elsewhere. This is not merely on account of the inevitable inaccuracies. Much of the text given here *is* accurate; that is, Wittgenstein did say the words in the text or something very close. But he did not read the material; he did not correct it; he was not in a position to throw any of it away. *Much* here he would have discarded. In fact he often did point out in the lectures that something he had said was misleadingly put; he had no opportunity to say that of any of the rest.

I am very grateful to Yorick Smythies, who kindly lent his notes to Rush Rhees for use in preparing this volume, to Norman Malcolm, for allowing a copy of his notes to be used, and to Mrs. Mildred E. Bosanquet and the late Mr. G. C. Bosanquet, who gave permission for the use of the notes taken by their son R. G. Bosanquet. I am more than grateful to Rush Rhees, without whom the volume would not have been possible at all. The idea of publishing a version of the 1939 notes was his: he thought that if a text *could* be put together from the different versions, it might usefully be included with some earlier material in a single volume on the foundations of mathematics. With that idea, he gave me his own notes to the lectures and, with them, the material he had obtained through the consent of Smythies, Malcolm, and the Bosanquets. He has done much to help in the preparation of this present volume, at first in connexion with the originally planned volume of which it was to be a part, and also later, after we had decided that the 1939 material should be published separately. I am especially grateful for his detailed comments on the entire manuscript; it was extremely important that the whole be checked by someone actually present at the lectures. His suggestions were invaluable and have saved me from numerous errors and infelicities.

I was helped in the preparation of this volume by a Summer Grant from the University of Virginia; a Small Grant from the university covered some incidental expenses. I am very grateful for this assistance.

CORA DIAMOND

Charlottesville, Virginia

Wittgenstein's Lectures on the Foundations of Mathematics

I

I am proposing to talk about the foundations of mathematics. An important problem arises from the subject itself: How can I—or anyone who is not a mathematician—talk about this? What right has a philosopher to talk about mathematics?

One might say: From what I have learned at school—my knowledge of elementary mathematics—I know something about what can be done in the higher branches of the subject. I can as a philosopher know that Professor Hardy can never get such-and-such a result or must get such-and-such a result. I can foresee something he must arrive at.—In fact, people who have talked about the foundations of mathematics have constantly been tempted to make prophecies—going ahead of what has already been done. As if they had a telescope with which they can't possibly reach the moon, but can see what is ahead of the mathematician who is flying there.

That is not what I am going to do at all. In fact, I am going to avoid it at all costs; it will be most important not to interfere with the mathematicians. I must not make a calculation and say, "That's the result; not what Turing says it is." Suppose it ever did happen—it would have nothing to do with the foundations of mathematics.

Again, one might think that I am going to give you, not new calculations but a new interpretation of these calculations. But I am not going to do that either. I *am* going to talk about the interpretation of mathematical symbols, but I will not give a new interpretation.

Mathematicians tend to think that interpretations of mathematical symbols are a lot of jaw—some kind of gas which surrounds the real process, the essential mathematical kernel.[1] A

1. Cf. G. H. Hardy, "Mathematical Proof", in *Mind* 38 (1929), 18: ". . . what Littlewood and I call *gas*, rhetorical flourishes designed to affect psychology,

philosopher provides gas, or decoration—like squiggles on the wall of a room.

I may occasionally produce new interpretations, not in order to suggest they are right, but in order to show that the old interpretation and the new are equally arbitrary. I will only invent a new interpretation to put side by side with an old one and say, "Here, choose, take your pick." I will only make gas to expel old gas.

I can as a philosopher talk about mathematics because I will only deal with puzzles which arise from the words of our ordinary everyday language, such as "proof", "number", "series", "order", etc.

Knowing our everyday language—this is one reason why I can talk about them. Another reason is that all the puzzles I will discuss can be exemplified by the most elementary mathematics—in calculations which we learn from ages six to fifteen, or in what we easily might have learned, for example, Cantor's proof.

Another idea might be that I was going to lecture on a particular branch of mathematics called "the foundations of mathematics". There is such a branch, dealt with in *Principia Mathematica,* etc. I am not going to lecture on this. I know nothing about it—I practically know only the first volume of *Principia Mathematica.*

But I will talk about the word "foundation" in the phrase "the foundations of mathematics". This is a most important word and will be one of the chief words we will deal with. This does not lead to an infinite hierarchy. Compare the fact that when we learn spelling we learn the spelling of the word "spelling" but we do not call that "spelling of the second order".

I said "words of ordinary everyday language". Puzzles may arise out of words not ordinary and everyday—technical mathematical terms. These misunderstandings don't concern me.

---

pictures on the board in the lecture, devices to stimulate the imagination of pupils." Cf. also J. E. Littlewood, *Elements of the Theory of Real Functions* (Cambridge, 1926), p. vi.

They don't have the characteristic we are particularly interested in. They are not so tenacious, or difficult to get rid of.

Now you might think there is an easy way out—that misunderstandings about words could be got rid of by substituting new words for the old ones which were misunderstood. But it is not so simple as this. Though misunderstandings may sometimes be cleared up in this way.

What kind of misunderstandings am I talking about? They arise from a tendency to assimilate to each other expressions which have very different functions in the language. We use the word "number" in all sorts of different cases, guided by a certain analogy. We try to talk of very different things by means of the same schema. This is partly a matter of economy; and, like primitive peoples, we are much more inclined to say, "All these things, though looking different, are really the same" than we are to say, "All these things, though looking the same, are really different." Hence I will have to stress the differences between things, where ordinarily the similarities are stressed, though this, too, can lead to misunderstandings.

There is one kind of misunderstanding which is comparatively harmless. For instance, many intelligent people were shocked when the expression "imaginary numbers" was introduced. They said that clearly there could not be such things as numbers which are imaginary; and when it was explained to them that "imaginary" was not being used in its ordinary sense, but that the phrase "imaginary numbers" was used in order to join up this new calculus with the old calculus of numbers, then the misunderstanding was removed and they were contented.

It is a harmless misunderstanding because the interest of mathematicians or physicists has nothing to do with the 'imaginary' character of the numbers. What they are chiefly interested in is a particular technique or calculus. The interest of this calculus lies in many different things. One of the chief of these is the practical application of it—the application to physics.

Take the case of the construction of the regular pentagon. Part of the interest in the mathematical proof was that if I draw a circle and construct a pentagon inside it in the way prescribed, a regular pentagon as *measured* is the result under normal circumstances.—And of course the same mathematical statement may have a number of different applications.

Another interest of the calculus is aesthetic; some mathematicians get an aesthetic pleasure from their work. People like to make certain transformations.

You smoke cigarettes every now and then and work. But if you said your work was smoking cigarettes, the whole picture would be different.

There is a kind of misunderstanding which has a kind of charm:

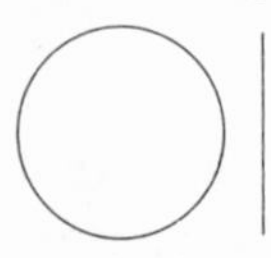

"The line cuts the circle but in imaginary points." This has a certain charm, now only for schoolboys and not for those whose whole work is mathematical.

"Cut" has the ordinary meaning: ∅ . But we prove that a line always cuts a circle—even when it doesn't. Here we use the word "cut" in a way it was not used before. We call both "cutting"—and add a certain clause: "cutting in imaginary points, as well as real points". Such a clause stresses a likeness.—This is an example of the assimilation to each other of two expressions.

The kind of misunderstanding arising from this assimilation is not important. The proof has a certain charm if you like that kind of thing; but that is irrelevant. The fact that it has this charm is a very minor point and is not the reason why those calculations were made.—That is colossally important. The calculations here have their use not in charm but in their practical consequences.

It is quite different if the main or sole interest is this charm—if the whole interest is showing that a line does cut when it doesn't, which sets the whole mind in a whirl, and gives the pleasant feeling of paradox. If you can show there are numbers bigger than the infinite, your head whirls. This may be the chief reason this was invented.

The misunderstandings we are going to deal with are misunderstandings without which the calculus would never have been invented, being of no other use, where the interest is centered entirely on the words which accompany the piece of mathematics you make.—This is not the case with the proof that a line always cuts a circle. The calculation becomes of no less interest if you don't use the word "cut" or "intersect", or not essentially.

Suppose Professor Hardy came to me and said, "Wittgenstein, I've made a great discovery. I've found that . . ." I would say, "I am not a mathematician, and therefore I won't be surprised at what you say. For I cannot know what you mean until I know how you've found it." We have no right to be surprised at what he tells us. For although he speaks English, yet the meaning of what he says depends upon the calculations he has made.

Similarly, suppose that a physicist says, "I have at last discovered how to see what people look like in the dark—which no one had ever before known."—Suppose Lewy says he is very surprised. I would say, "Lewy, don't be surprised", which would be to say, "Don't talk bosh."

Suppose he goes on to explain that he has discovered how to photograph by infra-red rays. Then you have a right to be surprised if you feel like it, but about something entirely different. It is a different kind of surprise. Before, you felt a kind of mental whirl, like the case of the line cutting the circle—which whirl is a sign you haven't understood something. You shouldn't just gape at him; you should say, "I don't know what you're talking about."

He may say, "Don't you understand English? Don't you understand 'look like', 'in the dark', etc.?" Suppose he shows you some infra-red photographs and says, "This is what you look like in the dark." This way of expressing what he has discovered is sensational, and therefore fishy. It makes it look like a different *kind* of discovery.

Suppose one physicist discovered infra-red photography and another discovered how to say, "This is a portrait of someone in the dark." Discoveries like this have been made.

I wish to say that there is no sharp line at all between the cases where you would say, "I don't know at all what you're talking about" and cases where you would say, "Oh, really?" If I'm told that Mr. Smith flew to the North Pole and found tulips all around, no one would say I didn't know what this meant. Whereas in the case of Hardy I had to know how.—In the case of the dark he only got an impression of something very surprising and baffling.

There is a difference in degree.—There is an investigation where you find whether an expression is nearer to "Oh, really?" or "I don't yet . . ."

Some of you are connected on the telephone and some are not.—Suppose that every house in Cambridge has a receiver but in some the wires are not connected with the power station. We might say, "Every house has a telephone, but some are dead and some are alive."—Suppose every house has a telephone case, but some cases are empty. We say with more and more hesitation, "Every house has a telephone." What if some houses have only a stand with a number on it? Would we still say, "Every house has a telephone"?

Suppose Smith tells the municipal authorities, "I have provided all Cambridge with telephones—but some are invisible." He uses the phrase "Turing has an invisible telephone" instead of "Turing has no telephone".

There is a difference of degree. In each case he has done something but not the whole. As he does less and less, in the end what he has done is to change his phraseology and nothing else at all.

Suppose we said, there being only a difference of degree, "Smith has provided all Cambridge with telephones." If such a difference is allowed, couldn't one say, "How did you come by this? I don't yet know what you mean."

We learn our ordinary everyday language; certain words are taught us by showing us things, etc.—and in connexion with them we conjure up certain pictures. We can then change the use of words gradually; and the more we change it, the less appropriate the picture becomes, until finally it becomes quite ridiculous. In the earlier cases we should say Smith was exaggerating or using high-flown language; finally we should say that he was simply using sophistry to cheat us.

To think this difference is irrelevant because it is a difference of degree is stupid.

This can only be said to confuse yourself or cheat yourself. If you do say it, it is only because you like to say you have provided the whole of Cambridge with telephones.

To understand a phrase, we might say, is to understand its use.

Suppose a man says that he has flown to the North Pole and has seen tulips there; and it turns out he means he saw there certain vortices of air and cloud which looked like tulips from his airplane. He says, "You mustn't think these tulips grow. They can only be seen from above. No seed, etc."—Here he is cheating in his use of this word. We should say we hadn't understood him. And if he was in the habit of saying this sort of thing, we should have the right, when he told us something which seemed surprising, to say to him, "I do not know what you mean. Tell me exactly what you mean, or else I may be cheated."

If a man says, "I flew to the North Pole", then one immediately thinks that one knows a lot about it, for example, that he crossed the Arctic Circle, etc.—If he said, "In my way of flying this doesn't hold", and he has been in a Cambridge laboratory all the time—he has described a new scientific process in old words, and we would say we didn't understand him. The picture he makes does not lead us on.

How much do we know of what he's talking about? By the words of ordinary language we conjure up a familiar picture—but we need more than the right picture, we need to know how it is used.

Suppose I said, "This is a picture of Moore.

It's an exact picture, but in a new projection. You mustn't think . . ."—If I say, "This is a picture of him", it immediately suggests a certain way of usage. For instance, I might say, "Go and meet So-and-so at the station; you will know him because this is a picture of him." Then you may take the picture and use it to find him. But you couldn't do the same with my picture of Moore. You don't understand my picture of Moore because you don't know how to use it.

Similarly, you only understand an expression when you know how to use it, although it may conjure up a picture, or perhaps you draw it.[2]

An expression has any amount of uses. How, if I tell you a word, can you have the use in your mind in an instant? You don't. You may have in your mind a certain picture or pictures, and a piece of the application, a representative piece. The rest can come if you like.

What is a 'representative piece of the application'? Take the following example. Suppose I say to Turing, "This is the Greek letter sigma", pointing to the sign σ. Then when I say, "Show me a Greek sigma in this book", he cuts out the sign I showed him and puts it in the book.—Actually these things don't happen. These misunderstandings only immensely rarely arise—although my words might have been taken either way. This is because we have all been trained from childhood to use such phrases as "This is the letter so-and-so" in one way rather than another.

When I said to Turing, "This is the Greek sigma", did he get the wrong picture? No, he got the right picture. But he didn't understand the application.

Similarly if I say to Lewy, "What is a Greek sigma?" and Lewy writes σ in the corner of the blackboard, then we say that Lewy knows what a sigma is. But it might turn out that he thought that the sign was only a sigma when written in the corner of the blackboard—perhaps because his schoolmaster wrote it there or something of the sort. Then we should say that after all he did not understand.—Or he draws sigmas like this: [3]

σ σ σ σ σ δ δ

He had the right picture in his mind, namely a picture of the sign σ; but he put it to the wrong use.

I say

2. (From "Suppose I said".) The versions of this passage in B and S are quite different. The text here is based on both; it could have been done very differently.

3. B's version of the second way Lewy goes wrong is entirely different. The text here is based on what is given only very sketchily in S.

(1) He understands it if he always uses it right in ordinary everyday life, millions of times.

(2) If he does this [Wittgenstein drew a sigma], we take his doing this as a criterion of his having understood.

Because in innumerable cases it is enough to give a picture or a section of the use, we are justified in using this as a criterion of understanding, not making further tests, etc.

I will be concerned with cases where having a picture is no guarantee whatever for going on in the normal way.

I will be concerned with cases where the use of words has been distorted gradually, so that a man points to a picture and then doesn't go on in anything like the ordinary way. So we don't know whether to say he has been to the North Pole or that we don't understand what he means.[4]

We will come to cases where I will point to a statement and say, "Is this similar to nonsense or to something that is surprising?"

I may be inclined to say, "Surely this is nonsense." You might say, "Isn't this arrogance? Shouldn't we say, 'Aren't you inclined to call this nonsense?' or 'This is nearer to the kind of expression of which we say, "I don't know what you mean" than the kind of which we say, "I know what you mean but I don't know how it happened".' "

One German philosopher[5] talked about "the knife without a handle, the blade of which has been lost". Shall we say that this is nonsense? And when do we say that it is no longer correct usage of the word "knife" but is nonsensical usage?

Suppose in the case of the telephones, I say to Turing, "Is this nearer to the ordinary case from which the phrase 'providing people with telephones' is drawn, or is it nearer to the absurd case I constructed, the man who simply changes his phraseology?" By talking this out, I may attract a man's attention to the nearness of what he does to [the absurd case]. If it doesn't do,

4. This sentence is doubtful. The sentences in B and S from which it is constructed make slightly different points.

5. Lichtenberg.

I can say, "Well, if this is no use, then that is all I can do." If he says, "There isn't an analogy", then that is that.

This means that I will try to draw your attention to a certain investigation.

You might, to be very misleading, call this investigation an investigation into the meanings of certain words. But this is apt to lead to misunderstandings.

The investigation is to draw your attention to facts you know quite as well as I, but which you have forgotten, or at least which are not immediately in your field of vision. They will all be quite trivial facts. I won't say anything which anyone can dispute. Or if anyone does dispute it, I will let that point drop and pass on to say something else.

One talks of mathematical discoveries. I shall try again and again to show that what is called a mathematical discovery had much better be called a mathematical invention.

In some of the cases to which I point, you will perhaps be inclined to say, "Yes, they had better be called inventions"; in other cases you may perhaps be inclined to say, "Well, it is difficult to say whether in this case something has been discovered or invented."

II

There is a puzzle about what we mean by saying that we understand a phrase or symbol. This arises because there seem to be two different sorts of criteria for understanding. If I ask you, "Do you understand 'book', 'house', 'two'?" you will immediately say "Yes". And if I ask, "Are you sure?" you might say, "Of course I'm sure. Surely I must know whether I understand it or not." Yet on the other hand whether you do understand it will come out in the way you use it, when you say "This is a house", "This is a bigger house than that", etc.

If it is true that you can understand a symbol *now*, and that this means you can apply it properly—then, one is inclined to say, you must have the whole application in your mind.

It may be all in your mind: for example, a complete diagram, or a page with rules. I will [say], "Say what you like."

But suppose we had the page of rules in our mind—does that necessarily mean we'll apply the word rightly? Suppose we both had the same page of rules in our minds, would this guarantee that we both applied them alike? You may say, "No, he may apply them differently." Whatever goes on in his mind at a particular moment does not guarantee that he will apply the word in a certain way in three minutes' time.

Should we then say that a man can never know whether he understands a word? If we say this, where shall we stop? We can't even say, "We will know it as time goes on." Suppose there were six uses of the word "house", and I used it correctly in each of the six ways; is it clear I will use it correctly the next time?

The use of the word "understand" is based on the fact that in an enormous majority of cases when we have applied certain tests, we are able to predict that a man will use the word in question in certain ways. If this were not the case, there would be no point in our using the word "understand" at all.

Suppose you say, "What does it mean for a man to understand a sign?"—You might say, "It means he gets hold of a certain idea."

Then if two people—Lewy and I—get hold of the same idea of 'two', we both understand it in the same way.—Suppose he had got hold of the same idea of 'two' as I, whatever that means. What if he used it differently in future? Would I still say he has got hold of the same idea? You might say, "Yes, he's got hold of the same idea, but applies it differently."

Suppose someone said, "Couldn't there be telepathy, and I know that Lewy has got hold of the same idea as I have? Or a medium might tell us." Would we say it was understood in the same way if it was applied in different ways? In fact it is clear that under those circumstances whatever the medium saw or said would be irrelevant to the question.

'Having the same idea' is only interesting if (a) we have a criterion for having the same idea, (b) this guarantees that we use the word in the same way. In that case anything can be the same idea, e.g., a picture in the mind.

This definition of two:

"This is two" | |

is as good a definition as Russell's, as good a definition as there is in the world. It can be misinterpreted, but so what? So can all definitions.

We do say that there may be a 'flash of understanding'—this is puzzling. How can understanding come in a flash?

Suppose two people sit down and say, "Let's play chess." They have the intention of playing chess. But chess is defined by means of its rules. If you change even one rule it would be a different game.—Suppose I say, "How do you know you intend to play a game of chess? Do you know that you will follow all these rules? Do you have all these rules in your head now?"—Suppose you have a page with the rules in your head. How do you know that you will apply them rightly? You may say, "There will also be rules for the way these rules are applied." But will you have the application of these in your head?

Should you therefore say, "I believe that I intend to play chess, but I don't know. Let's see"?—just as Russell once suggested that we don't know what we wish, don't know whether we want an apple or not.[1]

Suppose we said, "What he said was just a description of his state of mind." But why should we call the state of mind he's in at present "intending to play chess"? For playing chess is an activity, an activity we all know.

One might say, " 'Intending to play chess' is a state of mind which experience has shown generally to precede playing chess." But this will not do at all. Do you have a peculiar feeling and say, "This is the queer feeling I have before playing chess. I wonder

1. *Analysis of Mind,* Lecture III.

whether I'm [going] to play?"—This queer feeling which precedes playing chess one would never call "intending to play chess".[2]

Well, how is one taught the meaning of the expression "I intend to play chess"? One sees that it is the sort of expression which people use when sitting down at a chess board; but of course they sometimes say it when not sitting down at a chess board. Yet saying this generally goes with certain actions and not with certain other actions. (Suppose I say, "I now intend to play chess" and then undress.) Similarly it often goes with having certain images; but of course one can have any images when intending to play chess.

There are cases where we should say, "I did intend to play chess when I said so, but a second later I didn't"—when, for example, I had walked out immediately after saying it. If someone asked me what I had meant, this could be said—exceptionally. They might think me slightly queer, but that is all. For it might have been the case that I had suddenly thought of something else which had to be done.—But if that were the rule instead of the exception, if there were a race of men who always walked straight out of the room whenever they said "I intend to play chess" —would we say that they used the phrase in the same way we do?

One might be puzzled about this. One might say, "If it can happen in one case, why not in all?"—A word has a use, a technique of usage. If I usually use it one way and just occasionally in another, then we can say that that case is an exception, but we cannot say this if I always use it in that other way.

I have been considering the word "intend" because it throws light on the words "understand" and "mean". The grammar of the three words is very similar; for in all three cases the words seem to apply both to what happens at one moment and to what happens in the future.

2. (From "Do you have".) This passage is based on B, M, and S, the material from B having been altered to make it compatible with the rest.

What is a momentary act of understanding?

Suppose that I write down a row of numbers

1 4 9 16

and say, "What series is this?" Lewy suddenly answers, "Now I know!"—It came to him in a flash what series it is.

Now what happened when he suddenly understood what series it was? Well, all sorts of things might have happened. For instance, the formula "$y = x^2$" might have come into his mind—or he might have pictured the next number. Or he might just have said, "Now I know!" and gone on correctly.

But suppose that the formula "$y = x^2$" had struck him. Does that guarantee that he will go on and continue the series in the right way? Well, in an overwhelming number of cases, yes; he will go on correctly.

But now suppose that he goes on all right until 100, and then he writes "20,000". I should say, "But that is not right. Look, you have not done to 100 the same as you did to 99 and all the previous numbers." But suppose he stuck to it and said that he had done the same thing with 100 as he had done with 99.

Now what is doing the same with 100?—One might put the point I want to make here by saying, "99 is different from 100 in any case; so how can we tell whether something we do to 99 is the same as something we do to 100?"

One might say, "It's clear what 'the same' means—it's utterly unambiguous. We have an absolutely unequivocal paradigm for 'the same'," [Wittgenstein held up a piece of chalk] "This is the same as this." One can say that everything is the same as itself.

It might seem as if, if I take two pieces of chalk and say, "This is bigger than this", then what I say might be ambiguous. For it would be quite consistent with that explanation of the phrase "bigger than" that it should mean, for instance, 'to the left of'. Similarly, if I had pointed in turn to two pieces of chalk and said, "This is the same as that", I might have meant that they were the same size or the same shape or the same colour or many other things. But to say that everything is the same as itself seems utterly unambiguous.

Yet suppose I say, "You don't know what it is to do the same to 100 as 99. Well, I'll show you. This is the same as this." But he might reply, "Very well, but how am I to apply that definition to this case?"

Suppose I say, "Every patch fits exactly with its background" instead of "Everything is the same as itself." "This chalkmark fits exactly into its surroundings."—Then I am talking as if there were a hole into which I had fitted the chalkmark, or as if the chalk were surrounded by a glass case into which it fitted.—But we can talk of a piece of ice fitting into a glass, not of water fitting into a glass.

"We have one sure paradigm of equality and that is the equality of a thing with itself."—The point is that this gets us no further.

If he does something different with 100 from what he did with 99, shall we say that he understands squaring in a different way from us or in the same way? Well, there are different cases.

He might differ from us systematically. For instance, every time he got to a power of 100 he might do something queer. In that case we might say that by "x^2" he means '. . .' (and here we write down a formula).

Similarly we might teach him to 'add two', and he might do it all right up to 100, but then after 100 he adds 3, then after 1000 4, and so on. In this case we might say he had misunderstood us systematically; and we might succeed in correcting this. But there is no sharp line between systematic and unsystematic misunderstandings.

Suppose I teach Lewy to square numbers by giving him a rule and working out examples. And suppose these examples are taken from the series of numbers from 1 to 1,000,000. We are then tempted to say, "We can never really know that he will not differ from us when squaring numbers over, say, 1,000,000,000. And that shows that you never know for sure that another person understands."

But the real difficulty is, how do you know that you yourself understand a symbol? Can you really know that you know how to square numbers? Can you prophesy how you'll square tomorrow?—I know about myself just what I know about him; namely that I have certain rules, that I have worked certain examples, that I have certain mental images, etc., etc. But if so, can I ever know if I have understood? Can I ever really know what *I* mean by the square of a number? because I don't know what I'll do tomorrow.

We are inclined to think of *meaning* as a queer kind of mental act which anticipates all future steps before we make them.

Suppose that, when Lewy writes 20,000 instead of 10,000, I say to him, "No, I didn't mean that when I taught you. I meant you to write 10,000." That doesn't mean that, while I was explaining the rule for squaring numbers to him, I was at the same time performing the mental act of 'intending him to write 10,000 and not 20,000'. For in all probability I was not thinking of those numbers at the time.

But to say, "I am sure I meant him to write 10,000 and not 20,000 when he came to square 100" is like saying, "I am sure that I should have jumped into the water if Arabella had fallen in."

Should one then say that if I write $y = x^2$, where x is to take all the integers, that it is not determined what is to happen at any particular point?

I might say, "What is it determined by?" By this ($y = x^2$) together with the examples which I work out and the rules I give for its application, or by the range of the exercise? [3]

There are two senses of "determine".

(1) The question "Does the formula determine a series?" may mean 'Do people trained in a certain way generally go on writing down a certain series? Do they act in the same way when confronted with this formula and asked to write down its series?'

(2) There is a sense of "determine" in which it determines a series, in the sense in which $y = \pm x^2$ or $y = z + x^2$ (where z is undetermined) do not determine a series.

Hence one can ask, "Does the formula determine a series?" and mean either "Do most people act in the same way in this connexion?" or "Is it a formula of this kind or that?" "The

3. This paragraph, based on B and S, is quite doubtful.

formula determines . . ." can be used as a description of the behaviour of people or a description of the formula.[4]

Does my pointing determine the way he goes?—Do people normally go in one way? Yes. Or we might have a convention by which we distinguish pointing which determines the way from pointing which does not.

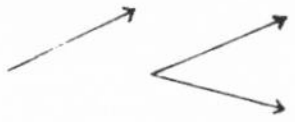

Pointing in the second way indicates that it does not matter in which of the directions one goes.

"Did your pointing determine the way he was to go?" might then mean "Did you point in one direction or in two?" [5]

Lewy: I want to say that we might by that question mean "Is it impossible for him to misunderstand me?"

Wittgenstein: But that comes to saying, "Is my pointing correctly understood whatever he does?" For instance, I might point in a certain direction and then say, "Good man, he has done what I told him to do", whether he just walks about the room or goes out or sits down or does anything else. Then the question, "Is it impossible for him to misunderstand me?" is [not] like saying, "Is there another interpretation? I can't think of one."

"Does the formula '$y = x^2$' determine what is to happen at the 100th step?"

This may mean, "Is there any rule about it?"—Suppose I gave you the training below 100. Do I mind what you do at 100? Perhaps not. We might say, "Below 100, you must do so-and-so. But from 100 on, you can do anything." This would be a different mathematics.

If it means, "Do most people after being taught to square numbers up to 100, do so-and-so when they get to 100?", it is a completely different question. The former is about the operations of mathematics but the latter is about people's behaviour.

4. Compare *Remarks on the Foundations of Mathematics,* Part I, §§1–3.

5. This paragraph was constructed using material from B, M, and S, the order being determined by S. This required some modification of the material from the other two.

Suppose someone said, "We know by an intuition what to do when we get to 100." Then one will also have to have an intuition to know how to continue the series 2, 2, 2, 2, . . . , in order to insure that one does not continue it

2, 2, 2, 2, 3, 3, 3, 3, 4, 4, 4, 4, 5, 5, . . .

You must have an intuition at each step in order to know that the number you put down is the same as the preceding one.

What one means by 'intuition' is that one knows something immediately which others only know after long experience or after calculation. For instance, one might say, "Although I only saw Smith twice, I knew by intuition he was a brave man and would jump in if Jones fell into the water."

Did I know by intuition that Smith would jump in after him if he fell in—if Smith *didn't* jump in?

If Lewy knew without a calculation that 1365 × 79 = . . . , we'd say he knew it by intuition, that he had mathematical intuition. The proof of having known by intuition is something different from intuition: he puts down without calculation what we put down with calculation. And a man knows anatomy by intuition if he can pass the exam without studying, which we can only pass by studying.—If we *all* knew by intuition and by intuition alone, this isn't what we could possibly call intuition.

If he failed the exam we might say either that intuition might go wrong, and that he'd had an intuition but a wrong one, or that we might think we had had an intuition when we hadn't.

One might say that most of us only know that 25 × 25 = 625 by calculation but that a few know it by intuition, whereas we all know how to go on: 1, 2, 3, . . . , by intuition. But suppose an intuition to go on: 1, 2, 3, 4, was a wrong intuition or wasn't an intuition?—A man is only said to know by intuition that 25 × 25 = 625 if 625 is in fact the result which we all get by calculation. But a man is said to know 1 + 1 = 2 not because 2 is in fact the result which we reach by calculation—for what sort of calculation should we use?—but because he says with the rest of us that 1 + 1 = 2.

The real point is that whether he knows it or not is simply a question of whether he does it as we taught him; it's not a question of intuition at all.

Doing after | ||, |||; going from 1 to 2 to 3, etc.,—is more like an act of decision than of intuition. (But to say "It's a decision" won't help [so much] as: "We all do it the same way.") [6]

We have all been taught a technique of counting in arabic numerals. We have all of us learned to count—we have learned to construct one numeral after another. Now how many numerals have you learned to write down?

Turing: Well, if I were not here, I should say $\aleph_0$.

Wittgenstein: I entirely agree, but that answer shows something. There might be many answers to my question. For instance, someone might answer, "The number of numerals I have in fact written down." Or a finitist might say that one cannot learn to write down more numerals than one does in fact write down, and so might reply, "the number of numerals which I will *ever* write down". Or of course, one could reply "$\aleph_0$" as Turing did.

Now should we say, "How wonderful—to learn $\aleph_0$ numerals, and in so short a time! How clever we are!"?—Well, let us ask, "How did we learn to write $\aleph_0$ numerals?" And in order to answer this, it is illuminating to ask, "What would it be like to learn only 100,000 numerals?"

Well, it might be that whenever numerals of more than five figures cropped up in our calculations, they were thrown away and disregarded. Or that only the last five figures were counted as relevant and the rest thrown away.—The point is that the technique of learning $\aleph_0$ numerals is different from the technique of learning 100,000 numerals.

Take the biggest numeral which has ever been mentioned. What is the difference between learning a technique of counting numerals up to that numeral and learning a technique which did not end at that numeral?

Well, it might have been that one's teachers said, "This series has no end." But how did you know what that meant?

They might have said that and then when one reached the numeral six billion, they might say, "Well, now we have got here, I need hardly say . . ." and shrug their shoulders with a slight

6. This sentence is a combination of two quite different ones, from M and S. It is very much a guess.

laugh.—So how did you know what they meant? Simply from the way in which the series was treated.

I did not ask, "How many numerals *are* there?" This is immensely important. I asked a question about a human being, namely, "How many numerals did you learn to write down?" Turing answered "$\aleph_0$" and I agreed. In agreeing, I meant that that is the way in which the number $\aleph_0$ is used.

It does not mean that Turing has learned to write down an enormous number. $\aleph_0$ is not an enormous number.

The number of numerals Turing has written down is probably enormous. But that is irrelevant; the question I asked is quite different. To say that one has written down an enormous number of numerals is perfectly sensible, but to say that one has written down $\aleph_0$ numerals is nonsense.

III

We talked about whether in the series of natural numbers it was determined what we had to do. We saw that the word "determine" can be used in two different ways. One can ask, "Does my pointing determine him to go in a certain direction?" and mean by that question either "Will he (or most people) go in a certain direction when I point?" or "Is one trained in such a way that, when I point, it is correct to go in a certain direction and incorrect to go in other directions?"

We asked, "How many numerals were you taught to write?" Would you answer this in the same way as "How many numbers are there?"

I might write down a number so large that you cannot imagine any experiential criterion to determine whether you had written down that number of numerals. "I don't know what would incline me to say that I had written down such a number."—For instance, suppose that one wrote down numerals in rows to facilitate counting. Then one knows what it would look like to have written down a hundred numerals; one might have written five rows of twenty numerals each. But I might name a numeral so large that you simply cannot imagine what that number of numerals would look like. The physicist might say, "We have no way of measuring such numbers." You might say, "By the training I have got, I clearly haven't learned to write down as many numerals as that."

Now why did I ask, "How many numerals were you taught to write?" and not "How many cardinal numbers are there?" There is a great difference between the two because the first is not a mathematical question. I wanted a non-mathematical statement containing "$\aleph_0$".

Some propositions belong to mathematics but other propositions containing mathematical symbols are not mathematical propositions. One could say that mathematical propositions containing the numeral "2" are not about 2. For example "$2 + 2 = 4$" isn't about 2 in the sense in which "There are 2 people on the sofa" might be said to be about people, the sofa and 2.—This is crudely put but it will be explained later.

One might also put it crudely by saying that mathematical propositions containing a certain symbol are rules for the use of that symbol, and that these symbols can then be used in non-mathematical statements.

You might say, "Is that true? Can we use all mathematical symbols in non-mathematical propositions?" You might say that there are parts of mathematics which have no application at all. Or can we always construct an application?

Turing: Surely those mathematical symbols which do not usually occur in non-mathematical statements are generally abbreviations for other mathematical symbols which do ordinarily occur in ordinary life.

Wittgenstein: I entirely agree; but I'd say that the word "abbreviation" is exceedingly misleading. The idea that a definition is an abbreviation is misleading all through mathematics. Is "$p | q$" an abbreviation for "$\sim p . \sim q$" or is "$\sim p . \sim q$" an abbreviation for "$p | q$"?—A definition signifies a change in technique.

One might give an application of the symbol $\int$ by measuring the width of this piece of chalk and finding it to be a quarter inch and then saying, "The moment of inertia of this face is $\int$ . . .", giving a formula. Am I not giving you a piece of information about the chalk? This is part of our ordinary speech; it is no more a mathematical proposition than "You have on a number of shoes which satisfies the equation $x + 5 = 7$" is a mathematical proposition.

Watson: Isn't it queer to call that an application of "$\int$"? For instance, suppose one said, "There are in this room as many people as the number of moves which are needed for Black to checkmate White from such-and-such a position", would that be called an application of chess?

Wittgenstein: Why, certainly it would be. Indeed it might be the case that we discovered a fixed correlation between the number of moves needed to checkmate from certain positions and the number of, say, atoms in certain molecules. Then, in order to discover the number of atoms in such-and-such a molecule, it might be easiest to set the chessmen in such-and-such a position and play chess. That would certainly be an application of chess.

Lewy: Is "Professor Hardy believes that $\aleph_1 > \aleph_0$" a mathematical statement?

Wittgenstein: No. It is no more a mathematical statement than "Willie said that $7 \times 8 = 54$" is a mathematical statement.

Why should not the only application of the integral and differential calculus etc. be for patterns on wallpaper? Suppose they were invented just because people like a pattern of this kind. This would be a perfectly good application. And in any case all mathematical symbols can be used in propositions which do not belong to mathematics; for they can occur in propositions of the form "He said that $25 \times 25 = 625$" or "He wrote down the proof that so-and-so."

In mathematics we have propositions which contain the same symbols as, for example, "Write down the integral of . . .", etc., with the difference that when we have a mathematical proposition time doesn't enter into it and in the other it does. Now this is not a metaphysical statement.

Turing: Does time enter into "This proposition is difficult to prove"?

Wittgenstein: That statement can be used in a temporal or in a (to use a misleading phrase) timeless way.

It can be used timelessly, can be made into a mathematical proposition—if the difficulty of proving the proposition is measured by the length, the number of transformations, etc.

It may mean, "Such-and-such people in such-and-such circumstances have difficulty in proving it." You may ask, "When? After drinking wine?" etc.

Or it may mean, "The proof is very long, requires 60 transformations." Why is this statement timeless? Why cannot one say, "It requires many transformations *now*"?

Suppose someone said, "This proposition is difficult to prove *now;* it requires many transformations *now*"—meaning "It is unlikely that at present we can prove it in fewer transformations." This is something entirely different and is not a mathematical proposition.

But there is a temptation to say "This proof requires many transformations now" is a mathematical statement. And this temptation arises from the fact that it can also be used in a different way.

Compare: "It is difficult to win from such-and-such a beginning in chess *now.*"

(a) Now—before dinner. After dinner it may be easy to win.

(b) You might talk of the development of the rules of chess. The rules had altered so that it was difficult to win now. It would be a statement about the rules of chess now. "Chess-1939" would be the name of a game.—"It is difficult to win in chess-1939." "In chess-1939 it takes 8 moves . . ."

Compare: "Proof of this proposition—1939". If you say, "The proof requires many transformations now", this would mean that what we now call a proof requires many transformations. Then it is a mathematical statement.[1]

1. (From "Suppose someone".) Much of this passage is based on S. In B the order is very different.

"$21 \times 36 = 756$" What do you mean by the proof of this proposition? Do you mean this figure?

$$\begin{array}{r} 21 \\ \underline{36} \\ 126 \\ \underline{63} \\ 756 \end{array}$$

If mathematical symbols were used for wallpapers and I were a wallpaper manufacturer, I might order an apprentice to make a certain wallpaper by telling him to prove repeatedly that $21 \times 36 = 756$.

You might call this figure the proof that $21 \times 36 = 756$, and you might refuse to recognize any other proof. Why do we call this figure a proof?

Suppose I train the apprentices of wallpaper manufacturers so that they can produce perfect proofs of the most complicated theorems in higher mathematics, in fact so that if I say to one of them "Prove so-and-so" (where so-and-so is a mathematical proposition), he can always do it. And suppose that they are so unintelligent that they cannot make the simplest practical calculations. They can't figure out if one plum costs so-and-so, how much do six plums cost, or what change you should get from a shilling for a twopenny bar of chocolate.—Would you say that they had learnt mathematics or not?

They know all the calculations but not their application. So one might say, "They have been taught pure mathematics."

They would use the words "proof", "equals", "more", etc., in connexion with their wallpaper designs, but it would never be clear why they used them. For these words are used in ordinary language. They could write down

$$\begin{array}{r} 21 \\ \underline{36} \\ 126 \\ \underline{63} \\ 756 \end{array}$$

and call it a proof. But if it were said, "The proof of Lewy's guilt is that he was at the scene of the crime with a pistol in his hand"—what is the connexion between this and calling the pattern a proof? They wouldn't know why it was called a proof.

Making wallpaper is an application and a most important one. But there are no other implications. It won't be clear what the connexion is between the way I apply these words to the wallpaper designs and the way they are applied in ordinary life.

Turing: The ordinary meanings of words like "three" will come out to some extent if they are able to do simple things like counting the number of symbols in a line.

Wittgenstein: It might come out to some extent.—But we talk of '+1' and talk of this being a number; and the ordinary use of "number" may have no connexion.[2]

What is the similarity between the figure and our proving in ordinary life that so-and-so is guilty?

[Turing gave an example of making squares and counting.]

"If you construct a rectangle 36 squares long and 21 squares wide, and count the squares in the ordinary way, you will reach this numeral."[3] The figure might be said to show this.—What about squares of astronomical proportions, where we get different results ordinarily? We might use numbers of different proofs.

If the figure is to be called a proof, we must be able to reproduce it always the same. For instance, if the figure on the bottom line were constantly changing, it would be useless to us as a proof.

Suppose that when we counted the squares we always got different results. Would the figure still be called a proof?

Turing: It would be a bad proof.

Wittgenstein: A bad proof of *what?* It would not prove what result we should get if we counted the squares on a certain occasion.

2. (From Turing's remark.) Both S and B have versions of this passage; but only S ascribes the suggestion to Turing. Wittgenstein's reply is based on both versions and may be inaccurate.

3. It is impossible to be certain what was said by Turing and what belongs to Wittgenstein's reply.

In fact the figure would be useless for physics. What would it prove? Nothing.

We call these things proofs because of certain applications; and if we couldn't use them for predicting, couldn't apply them, etc., we wouldn't call them proofs.—The word "proof" is taken from ordinary everyday language, and it is only used because the thing proves something in the ordinary sense.

Would the figure then cease to be a piece of arithmetic? It is what it is. You learned a certain technique when you were taught arithmetic, a way of writing things down. You could still carry on with that technique.

I could use the figure in order to predict what a man who has had the ordinary education will write down when asked what 21 times 36 is equal to. What will be unclear is why we use here the word "times" or "equals"—why we call these equal when ordinarily we call these heights equal. They seem in such a case to have no connexion with our ordinary life. But the use of these words is now justified by the application of mathematical calculations.

If multiplication were a complicated process, we might find it easier to count squares. And then we might conversely call the counting of the squares the proof that Professor Hardy will get 756 if he multiplies 21 by 36.

By the proof that $21 \times 36 = 756$ I could mean just this pattern. You may say, "Not only that pattern, but any pattern which proves that proposition, for example:

$$\begin{array}{r} 36 \\ \underline{21} \\ 36 \\ \underline{72} \\ 756 \end{array} \quad ."$$

But what is meant by 'any pattern which proves that proposition'? What is meant by calling patterns 'proofs'?—That we allow several patterns to be called proofs of the same proposition is due to the application of the symbols in question. Apart from their application we should not call any of the patterns 'proofs'. And under some circumstances, with certain applications we might call one of these two patterns a proof and the other not.

There is no 'general proof'.[4] The word "proof" changes its meaning, just as the word "chess" changes its meaning. By the word "chess" one can mean the game which is defined by the present rules of chess or the game as it has been played for centuries past with varying rules.

We fix whether there is to be only one proof of a certain proposition, or two proofs, or many proofs. For everything depends on what we call a proof.

It is not the case that there are two facts—the physical fact that if one counts the squares one gets 756 and the mathematical fact that 21 times 36 equals 756.

What I am out to show you could be expressed very crudely as "If you want to know what has been proved, look at the proof" or "You can't know what has been proved until you know what is called a proof of it."—But these are like exaggerations, partly true and partly false.

IV

Let's suppose a tribe which liked to decorate their walls with calculations. (An analogy with music.) They learn a calculus like our mathematics in school, but they do the calculations much more slowly than we do—not in a slapdash way. They never write the sign $\int$ without decorating it very carefully with different col-

4. Cf. *Remarks on the Foundations of Mathematics,* Part IV, §40.

ours. And they use the calculus solely for the purpose of decorating walls.

Suppose that I visit this tribe, and I want to anticipate what they're going to write. I find out the differential calculus and write it down in a very slapdash way, quickly—and find, "Oh yes, he's going to write down $\frac{x^3}{3}$." I would use my calculations to make a forecast of what they are going to write.

Suppose I invented these operations to make these forecasts. Would I be doing mathematics or physics? Would my results be propositions of mathematics or physics?

To put it another way: Suppose a people who learn to multiply solely in order to predict weights. They put measuring rods against the sides of parallelepipeds, read off the lengths on the measuring rods, and multiply—and say that that is the number of grams which will balance the object when put on the scales. They use multiplication only for this purpose and are in other respects so ignorant that they cannot add, divide, or perform any other mathematical calculations; suppose that they cannot even count.

In what we do we are always isolating calculuses.

Have they learnt mathematics or physics? Is the result a proposition of mathematics or physics?

One might say it is a proposition of physics because it is used to make predictions; it tells us what the weight will be.

But why should one not say "Both"?—They are making a calculation. But one thing is missing, a mathematical proposition. They can multiply and never say a mathematical proposition.

There is an idea that mathematics consists of propositions, whereas you might say it consists of calculations, which is a very different thing.

Is the proposition "$25 \times 25 = 625$", as a result of calculation, a proposition of mathematics or of physics? What would be the difference?

One might say that the answer to that question depends upon what they mean by the proposition. But it isn't clear what is meant by the phrase "what they mean by the proposition" nor what the criterion for it is. It would be easier to say it depends upon how they use the proposition. What would be the difference in the way it was used that would make it one or the other? [1]

Lewy: Wouldn't the difference lie in the sort of circumstance which would persuade them to say that their statement was wrong?

Turing: Wouldn't it be a proposition of mathematics if I said to somebody that I had performed a calculation and got this result?

Wittgenstein: I will have to talk a great deal about that word "got". There is a temporal 'got' (for example, "I got today but not yesterday") and a timeless 'got'.—In mathematical propositions, the 'is' is not temporal. It is absurd to say, "6 X 6 is 36 at 3 o'clock."

The point is that the proposition "25 X 25 = 625" may be true in two senses. If I calculate a weight with it, I can use it in two different ways.

First, when used as a prediction of what something will weigh —in this case it may be true or false, and is an experiential proposition. I will call it wrong if the object in question is not found to weigh 625 grams when put in the balance.

In another sense, the proposition is correct if calculation shows this—if it can be proved—if multiplication of 25 by 25 gives 625 according to certain rules.

It may be correct in one way and incorrect in the other, and vice versa.

It is of course in the second way that we ordinarily use the statement that 25 X 25 = 625. We make its correctness or incorrectness independent of experience. In one sense it is independent of experience, in one sense not.

Independent of experience because nothing which happens will ever make us call it false or give it up.

1. (From "One might say it".) The order of the material in this passage is quite uncertain. In particular, the first two sentences may belong at the end of the passage or closer to it.—Towards the end of the passage, Wittgenstein apparently referred to J. B. S. Haldane and the Albert Memorial.

Dependent on experience because you wouldn't use this calculation if things were different. The proof of it is only called a proof because it gives results which are useful in experience.

Suppose that we found that all ordinary objects had lengths which were multiples of the length of this piece of chalk. Then nothing would be more natural than to choose this chalk as our unit of length. Our unit of length is in that case dependent upon experience, in the sense that it is experience which makes us choose it. But if we later came across objects whose lengths were not multiples of this piece of chalk, we should not give up that unit of length.

One might say, "This piece of chalk is the unit of length", and mean that all objects have lengths which are multiples of this. In this case it is not independent of experience; it is an experiential proposition. But one can use the sentence "This piece of chalk is the unit of length" in quite a different way, in order to say something about the way one is going to measure lengths.

Turing: In the latter case isn't it like a definition?

Wittgenstein: But what is the right-hand side and the left-hand side of the definition?

One might write "This piece of chalk = 1 W df." But although one writes down the words and puts the equality sign and so on, one does not write down how the definition is to be used.

The important thing would be teaching a man the technique of measuring with this unit. It is no good simply writing "This piece of chalk = 1 W df." One has to say, "Take this piece of chalk and put it alongside the object in question, and then . . .", teaching him the technique.

We often put rules in the form of definitions. But the important question is always how these expressions are used.

Suppose someone knew logic but not mathematics. Could we teach him to multiply simply by definitions? Can the decimal system be taught by definitions?—If Russell can do all mathematics in *Principia Mathematica,* he ought to be able to work out $25^2 = 625$. But can he? How could decimal numbers be introduced into *Principia Mathematica*?—Russell and Frege said that by introducing some more definitions into their systems they could prove such things as $25^2 = 625$. But we cannot teach anybody to multiply by definitions.

Mathematics and logic are two different techniques. The definitions are not mere abbreviations; they are transitions from one technique to another, projections from one technique into another. They connect two different techniques.

Turing: Wouldn't Russell be able to prove something corresponding to $25^2 = 625$ in his symbolism?

Wittgenstein: Yes; but it is that phrase "corresponds to" which is the whole point. What does that mean? Isn't a piece of mathematics needed to show that the two do correspond?

It is immensely important to realize that definitions join two quite different techniques. Sometimes the difference is important, and sometimes it is trivial, as when we write "c" instead of "$a \times b$". But the fact that the difference is trivial should not blind us to the fact that these are two different techniques. Why should we say that "c" is merely an abbreviation of "$a \times b$"?

Are we to say that this

is merely an abbreviation of your face? It is in a projection different from the usual ones of pictures of your face.—And definitions are projections.

But I will come back to the question of definitions, and treat of it at some length, as it is of immense importance.[2]

All the calculi in mathematics have been invented to suit experience and then made independent of experience.

Suppose we observed that all stars move in circles. Then "All stars move in circles" is an experiential proposition, a proposition of physics.—Suppose we later find out they are not quite circles. We might say then, "All stars move in circles with deviations" or "All stars move in circles with small deviations."

2. There are some further remarks about definition in Lectures XXVII and XXVIII.

The simplest method of describing their paths might be to describe their deviations from the circles. Suppose I now say, "All bodies move in circles with deviations", meaning ——— is a circle with deviations—now I am no longer making a statement of physics. It is now a proposition of geometry; I have made it independent of experience.[3] I have laid down a proposition which provides a form of representation, a method of description—just as I did with the statement that this piece of chalk is the unit of length. It was this which made Turing say that the unit of length was introduced by a definition. But instead of saying that we could say: The statement becomes a rule of expression.

It is the same with "$25 \times 25 = 625$." It was first introduced because of experience. But now we have made it independent of experience; it is a rule of expression for talking about our experiences. We say, "The body must have got heavier" or "It deviates from the calculated weight."

I am trying to conduct you on tours in a certain country. I will try to show that the philosophical difficulties which arise in mathematics as elsewhere arise because we find ourselves in a strange town and do not know our way. So we must learn the topography by going from one place in the town to another, and from there to another, and so on. And one must do this so often that one knows one's way, either immediately or pretty soon after looking around a bit, wherever one may be set down.

This is an extremely good simile. In order to be a good guide, one should show people the main streets first. But I am an extremely bad guide, and am apt to be led astray by little places of interest, and to dash down side streets before I have shown you the main streets.

The difficulty of philosophy is to find one's way about. The real difficulty in philosophy is a matter of memory—memory of a peculiar sort.—A good guide will take one down each road a hundred times. And just as a guide will show one new streets every day, so I will show you new words.

3. (From "The simplest".) This material could have been put together in quite different ways.

I will sometimes have to give a whole lot of examples and then pull them together.

Here is an example: "Smith drew the construction of a pentagon." That is not a proposition of geometry. It is an experiential proposition and may be either true or false. But what about "Smith drew the construction of a heptagon"?

Turing: That is undoubtedly false.

Wittgenstein: Well, is it? Isn't it queer that the case of the pentagon is so different from the case of the heptagon?

Lewy: One might say that the phrase "construction of the heptagon" doesn't signify anything.

Turing: There is something queer about saying that "Smith drew the construction of the heptagon" is certainly false. For it suggests that it might be true but is certainly false.

Wittgenstein: Yes.—And if one says that it is certainly false it seems as though one ought to say that "Smith drew the construction of the pentagon" is certainly true. For it seems as though one is using the phrase "Smith drew the construction of the . . ." to mean "It is possible to construct the . . ."

"It is impossible to draw the construction of the heptagon; so the proposition that Smith drew it is false."—Isn't this queer? Compare it with "It's impossible to draw a heptagon; so the proposition that he drew it is false." This would probably be taken to mean something like "He has not got a ruler."

How do you know that it is impossible in Turing's sense to construct a heptagon with ruler and compass? You have proved it. But what if one were to say, "Well, it's pretty difficult"?

Lewy: I should reply that there is nothing which I should describe as 'constructing a heptagon with ruler and compasses'.

Wittgenstein: Yes—but that is a question of definition, and Turing did not treat it as a question of definition.

What does it mean to construct a heptagon with ruler and compasses? Well, using compasses is doing this sort of thing (here one performs a process which we all know); and similarly using a ruler is something which we all know. But how do you know that by doing this sort of thing one cannot construct a heptagon whose sides and angles, when measured, are found to be equal?

Suppose one asked, "How do you know that that pentagon on the board is regular?" Well, one answer might be, "Because I measured the angles and sides with a protractor and ruler, and I found them to be equal." But could not another sort of answer be given? We have another way of measuring the regularity of pentagons, and that is by drawing the construction. This second process turns the construction into a form of measurement. Why do we allow it as a form of measurement? Because it works: that is, because the two ways of measuring the regularity give the same result.

Why do we give the name "construction of the pentagon" to the process which we do give it to? Because the result which it in practice gives is a regular pentagon.

Turing: There are other reasons too.

Wittgenstein: Why yes, it is true that that is not the only reason. But if these 'other reasons' applied to the process, and yet the process did not give a regular pentagon, should we call it the construction of the pentagon?

"Smith drew the construction of a heptagon." We heard that this was a false proposition, because nobody can draw the construction of a heptagon. But might he not use the ruler and compass and have a heptagon result? Of course.—Of course heptagons hardly ever are produced that way. And though that fact does not entail that it is impossible to construct a regular heptagon with ruler and compasses, it is not altogether irrelevant. It is relevant in the way that mathematics is always dependent on its application.

We say that we cannot mathematically construct a heptagon. But if we find that we can construct a heptagon, do we give up the proposition that we cannot mathematically construct a heptagon? Of course not.—What then do we mean by saying that?

Turing: We mean that we cannot give instructions for constructing a regular heptagon.

Wittgenstein: But I might give instructions to someone and he might go on constructing heptagon after heptagon. Of course one might then say that he was not following the instructions. But what do you call 'following the instructions'?

Isn't it true that we have arranged our notation in such a way that there isn't anything which we would call the construction of a heptagon?

All that the mathematical proof that a regular heptagon cannot be constructed with ruler and compasses achieves is to give us good grounds for excluding the phrase "construction of the heptagon" from our notation. Hence "Smith drew the construction of the heptagon" is not false but meaningless. It uses an expression which has not only not been given a meaning, but has been excluded. It has been excluded for experiential reasons, although the statement that it is impossible is not an experiential statement.

Could one not prove that it is possible to construct a regular heptagon with ruler and compasses? For instance, one might prove that it is possible, but only in an infinite number of steps and therefore humanly impossible.

To sum up, I have tried to show that the connexion between a mathematical proposition and its application is roughly that between a rule of expression and the expression itself in use. We choose such a rule of expression for a mass of reasons. For instance, the mathematical proof that it is impossible to construct a heptagon gives us good reasons for excluding the phrase "construction of the heptagon" from our notation. These reasons are very complicated; but they show that, if we did not exclude that phrase, we should get into difficulties not only about this but about many other things too.—We could multiply in some other way but it would not be convenient.

Next time I ought first to try to say better what I have said so badly today. But secondly, I will talk about mathematical propositions which seem to have their applications in themselves, in mathematics.[4]

For example, consider "Equations of the form $ax^2 + bx + c = 0$

4. Wittgenstein did not reach this subject until Lecture XII.

have two roots". Or "The number of real numbers is greater than the number of rational numbers".—One might think that the difference between "$2+2=4$" and these is that these seem to be not only mathematical propositions but also propositions about mathematics. It looks as if they already have their application inside mathematics, and one need not look for another application. They look not like rules but like experiential propositions. But I will try to show that these statements are [rules] in just the same way as "$2+2=4$".

One counts the roots of an equation; how then can the statement that certain equations have two roots be a rule? I will say that it introduces a new symbol into our calculations: the word "root". For we do not ordinarily calculate in mathematics with the word "root".

V

Consider Turing's assertion: "It is certainly false that Smith drew the construction of the heptagon." It seems queer that one can say this without knowing what Smith did. We can't say, "Of course he drew the construction of the pentagon."

We might ask, "What is it he didn't draw?" But there doesn't appear to be an answer to this. For in the case of Smith drawing or not drawing the construction of the pentagon, we can draw a figure on the blackboard, and say, "This is what Smith drew—or didn't draw." But what the mathematicians have proved is, if anything, that there is no such figure to which we can point in the case of the heptagon.

Suppose that an undergraduate says that Professor Hardy in his lecture drew the construction of the heptagon on the blackboard. One would immediately say that the undergraduate had made a mistake. But what kind of mistake would one suspect?

One might think that by "heptagon" he meant pentagon, and that when asked what it was that Professor Hardy drew, he would draw on a piece of paper the construction of the pentagon.—But suppose that the undergraduate, when asked that, drew with ruler and compasses a heptagon which was found to be regular when measured. We might then say he had made a mistake about how mathematicians use the phrase "construction of the heptagon".

One can prove that it is possible to construct a pentagon. But what sort of possibility has been proved?

Suppose that I say, "I will prove to you that I can open that safe." One can here distinguish between the end (the opening of the safe) and the means (the jemmy). For we all know what it is like to open a safe. And we can show what is meant by it by drawing a picture of it or by opening other doors and safes and so on.—But in the case of the pentagon, is the end the regular pentagon, or the construction of the regular pentagon?

Turing: It is the construction, since it would be no good producing a regular pentagon by a fluke.

Wittgenstein: Yes. But of course one must be able to explain what a regular pentagon is, in order to explain what this construction is going to be a construction of.

Suppose we wish to construct a regular pentagon; the pentagon is what we want and the construction is a means to that end. We might explain what a regular pentagon is either by showing one or by describing it as a pentagon with all its sides and angles equal. But then we shall have to say that what the proof shows is that we can construct a pentagon whose angles and sides, when measured with a footrule and protractor, are found to be equal. Used in this way, "We can construct a pentagon" is a proposition of physics. It is not a mathematical proposition but an experiential one.

But we might use this sentence in such a way that it expressed a mathematical proposition.—It is clear that this way of putting [what the proof shows] is enormously misleading. For this proof that a regular pentagon can be constructed is a very simple proof; but when we ask what it proves, no simple answer seems to be forthcoming.

Of course, I am not showing that the mathematicians are wrong. It is merely that there is a misunderstanding somewhere.

For we do not quite seem to know what it is that we are constructing.

It seems as if the regular pentagon which mathematicians tell us can be constructed is different from the ordinary pentagon. For Euclidean geometry talks about sides being *equal*—but it gives no criterion for determining when sides are equal. One might say that Euclidean geometry never teaches us how to measure.

In ordinary life we have all sorts of criteria for equality. Should we say that Euclidean geometry talks of equal length in a different sense? Does it call something equal which in ordinary life we don't call equal? If so, it is difficult to say what is the use of it. It would seem to be of little importance.

Similar considerations apply to such things as equal weight, equal colour, equal number, etc. Aren't there very different criteria for equality in all these cases? For instance, we say that two lots of things are of equal number if a normally educated man counts them and finds them equal, or if we can put strings across from each member of the one lot to one member of the other, etc., etc.

Then is Euclidean geometry going wrong when it talks of equal length and misses out the most important thing: the criterion of equality?

Prince: One might say that it leaves one to choose one's own criterion.

Wittgenstein: Yes—but doesn't it apparently (this word "apparently" will have to be explained later) say, "Such-and-such lengths are equal"? "And then it leaves one to choose one's own criterion of equality."—"Well, then it doesn't say very much!"

If it leaves one to choose one's own criterion, it doesn't in any normal sense say two lengths are equal.

It does not seem to be at all the same sort of statement as "This pencil is the same length as that." For here one clearly has to know the criterion of sameness. What is the relation between a proposition of Euclid which tells you two lengths are equal and an ordinary experiential proposition which tells you two lengths are equal? What would you call this relation?

Compare it to the following question.

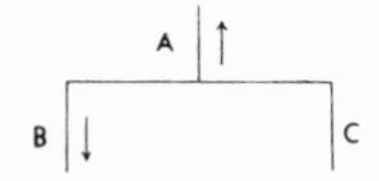

Suppose one wants to describe the motion of C. Where the motion of A doesn't come in, you might call it the normal case. Then describing the motion of B will describe the motion of C. But directly the motion of A comes in, this is not so. Then one might say that by describing the motion of B one has only given a way of describing the motion of C.

(How does Euclidean geometry manage to say anything about equal lengths, seeing that it never tells us how to measure? Well, given the *ordinary* methods of measurement, Euclidean geometry does describe what actually happens.)

If I say that B moves down with uniform velocity. I haven't yet described the motion of C. One might say I've given a partial description.—A partial descripton? One might reply that a partial description is no description at all and that it leaves us just where we were.

But are we left where we are? Not exactly, say if A is normally stationary or if it moves very slowly. If I begin to describe the motion of C by saying B moves down with velocity v, I may have given an almost exact description. I may have made it very easy to describe the motion of C (if, for instance, the movement of B is very complicated and that of A very simple). Or I may have described the normal motion of C (if, for instance, A is normally stationary).

If we give a law for the motion of B, we may have given the groundwork for a description of the motion of C. And sometimes the groundwork is very complicated, and when it is completed the rest is easy.

Now let us put the point in another way. Suppose the figure to be quite imaginary, and that all we have is a moving body C. The rest of it is only a model, which we may use to describe the movement of C—by saying how A and B move.

Describing the movement of B may be the groundwork of our

description. At first it was a description; we took A to be stationary. Then we said, "But you know, A is not quite stationary, only approximately so." Then one's description of B becomes only an approximate description of C; "It does so-and-so—pretty nearly." Finally A may begin to move around a lot. But we might say that the motion of B is the motion of C; A can move any damn how.—We have set up a standard of comparison—or a method of description, as in the star case. We are no longer giving a description.

In Euclidean geometry when one proves that two lengths are equal one is not saying that two lengths are equal in the ordinary sense; for one does not give a method of comparison. But one may provide a groundwork for the description of their comparative lengths. For it is in fact true that if we measure them by any ordinary method we shall find them to be equal in normal circumstances. But Euclidean geometry does not depend on experience; not giving the method of measurement makes it aloof.

Proving that the construction of the pentagon is possible is showing the construction of a pentagon. Compare: It is possible to bisect a line.

"I have shown you the construction of the bisection."

What have we said when we say that it is possible to construct a bisection? What is the possibility that we have shown? Are we to say that we have shown that this figure that we have drawn is possible? In what sense have we shown that it is possible?

Is the phrase "the bisection of a line" the proper name for this figure? What is the criterion for this being the construction of the bisection?

One might say that one will be able to prove from certain axioms that the line is bisected. But that will only mean that one will be able to construct from certain forms which we call 'axioms' the proposition that

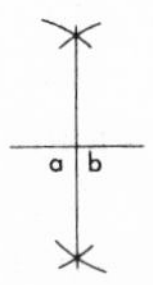

$a = b$. —This makes it look as if one has a general idea of what bisection is and then shows that this is a case of bisection.

But I am muddled now and cannot get clear at the moment what I want to say about this.

Take another example.—This is the solution of a certain puzzle:

Now what does this show? One might say that it proves that it is possible to fit the four pieces together. Or one might say that it proves that the rectangle consists of four parts of such-and-such a shape.—But which rectangle? The rectangle on the blackboard? But no one ever doubted that. Or not that particular rectangle but every rectangle? But how could it show that? Or *does* it show that lines can be drawn as in the figure in every rectangle? But who ever doubted that?—If you say this shows a possibility—a possibility of what?

Is a proof that a face can be drawn? Why is it that shows a possibility and not ?—But one can say that this shows something. For there might be a game in which one gives a child an iron ring, two sugar lumps, and two pieces of chalk, and asks him to make a face out of them. Then this figure proves that it is possible to do so.

As soon as I get into mathematics the means and the result become the same. But as soon as I distinguish between means and result, it is not mathematics.

For what does show the possibility of? It shows the possibility of putting the puzzle together. But is that a mathematical proposition or not?

One can say many things about that proposition. For instance, in what sense does the figure show that the puzzle can be put together? Does it prove that pieces of this size and shape can be put together? But they might dissolve when I tried to put them together; or they might be electrically charged so as to fly apart when I tried to get them near to one another; or they might even simply not fit.

When is it proved that they can be put together? When they are put together or before they are put together? But when they are put together, what does it show? And if they are not yet put together, is it not possible that they cannot be put together? It is like our old question, "Does it determine . . . ?" [1]

But we say that this shows that the puzzle can be put together because in an immense majority of cases you need only show this and the rest is simple. We always or almost always can put the puzzle together; it almost never happens that the pieces will dissolve, etc.

If we have such a figure we might say it shows that the puzzle can mathematically be put together, or that it can as far as geometry goes be put together.

Suppose we all try to put the bits together and can't; and we all say, "The bits can't be put together." Then someone draws this figure and then we say, "After all, it can be put together." What could we say he's done?

He hasn't done any putting together. He has given us a model. He has given us something which now makes it easy for us to put it together.—If we say, "This shows that it can be put together", we have given a new meaning to "can be put together"—different from the meaning we attached to this expression when we were scrambling around with the pieces; we have found a new criterion for it.

So it was absurd for me to ask, "Does this show that it can be

1. See Lecture II.

put together?" For of course it is possible that we shall not be able to put it together after he has drawn the picture; but if he draws the picture, then we say he has shown that the puzzle can be put together.

In the proof nothing has been put together. We call it a proof that the pieces can be put together because it is a picture of how these things can be put together. It is the paradigm of such things put together. And we can use it as a model.

Suppose I say, "It is possible to construct a pentagon." What is it that is possible? That peculiar figure? Here again you can say it is a model which can be used for constructing pentagons.

What a mathematician gives you is a model which can then be used for certain purposes.

I had wanted to start today from two points: (1) the question whether it is impossible to construct a heptagon, and (2) Lewy's answer the other day to a question of mine. Lewy said, "Well, I know what you want me to say." [2]

That was a severe criticism of me. For I have no right to want you to say anything except just one thing: "Let's see".—One cannot make a general formulation and say that I have the right to want to make you say that. For what could that general formulation be? My opinion? But obviously the whole point is that I must not have an opinion.

The only thing which I have a right to want to make you say is, "Let's investigate whether so-and-so is the case."

For instance, I have no right to want you to say that mathematical propositions are rules of grammar. I only have the right to say to you, "Investigate whether mathematical propositions are not rules of expression, paradigms—propositions dependent on experience but made independent of it. Ask whether mathematical propositions are not made paradigms or objects of comparison in this way." [3] Paradigms and objects of comparison can only be called useful or useless, like the choice of the unit of measure-

2. There is no record of this remark.

3. (From "I only have".) This passage has been put together from the two somewhat different versions in B and M; it could have been done in various ways.

ment. Similarly one can say that the construction of the pentagon is useful or useless.

If we prove that a certain mathematical proposition is not provable, then we may be said to be asserting a proposition of geometry; it is like asserting that the heptagon cannot be constructed. If we really prove that the heptagon cannot be constructed, it should be a proof which makes us give up trying—which is an empirical affair. And similarly with proving that a certain proposition is not provable.

It has been proved impossible to trisect an angle with ruler and compasses. If I do manage to trisect an angle thus, it must be easy for us to point out the mistake in my trisection and to say, "It was a fluke."

Whether or not we say, "There must be a mistake in the construction", is a question of decision. The proof that the trisection is impossible must give us good grounds for saying, "There must be a mistake." We must believe the proof rather than the trisection.

VI

How do we show that a line *can* be bisected?

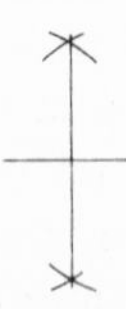

How does this show what can be done?—unless it means ". . . if you are sufficiently intelligent".

Do we prove that we can divide a line into two equal parts —meaning 'equal' according to the ruler? Clearly not. This would transcend mathematics. It is not a mathematical but an experiential proposition.

It might be said what is meant is that there is a system by which it can be proved from certain axioms, this length = this length.—But this only means, from Euclid's axioms there follows a proposition, so-and-so equals so-and-so.

So we must see what these axioms of Euclid's are.

Take the axiom: Between any two points a straight line can be drawn. This is obviously untrue, you might say. Try to draw a line between a point on the moon and a point on Sirius, whatever a point may be.—But clearly that is not what is meant by "can be drawn".

One may mean many different things by "can".

"Can you go to the chemist's and get something to gargle with?"—"I can't."

meaning—I have no legs
—I'm a lady
—I haven't time
—etc., etc.

You might say, "I can as far as my legs are concerned, but I can't as far as time is concerned." Or "I can go in respect of my legs but not in respect of time."

In the same way, you might think that Euclid tells us that we can in a certain respect draw a straight line between any two points. You might say, "We can, as far as Euclid is concerned."

Now what is the respect in which we can? We can, in the sense in which we can't draw a straight line between any three points. —But how is it we can't do this? Isn't there something queer about this? Isn't it something we couldn't even *try?* Or try to try, etc.

If someone says, "I'll show you how to draw a straight line through them":

—how do we explain what can't be done?

Suppose we explain: This is a line going through one point  ; this is a line going through two points  ; and this is a line going through three points  . And suppose he still says, "Yes; and this is, too:  ."

We might say, "But that is a different use of 'through'" or "The two cases are not similar" or "This isn't analogous to this."

I would say this. But the point is not what I would say, but why I say it.

One might say, although this is not a good way of putting it, that the words "same", "similar", and "analogous" are each used in two different senses. (I will talk a lot about these words in this lecture, and whatever I say about one of them will apply to all of them.)

They can be used this way. I do jerks, and then say, "Do the same", "Do the analogous thing"—and he takes hold of my hands, or he is at a loss what he is to do. But in fact this doesn't happen, because we have learnt the technique of using the word "same".

Similarly one can show a child how to multiply 24 by 37, and 52 by 96, and then say to it, "Now multiply 113 by 44 analogously." The child may then do one of many things. If he can't justify his action, we should go through it again and again, until we converted him to doing the same as us. The only criterion for his multiplying 113 by 44 in a way analogous to the examples is his doing it in the way in which all of us, who have been trained in a certain way, would do it. If we find that he cannot be trained to do it the same as us, then we give him up as hopeless and say he is a lunatic.

"We taught him multiplication up to 100, and then he did the analogous thing." "I want you to draw something analogous to this."—What are we saying? [1]

But if we want to show the child that one process is analogous to another, we say, "Come now, this is not analogous to that, but this *is* analogous to these. Surely now, *these* two are analogous." "What is analogous to doing this? This? or this?" etc.—But what sort of thing are we saying now? Surely what? Are we describing anything now? Before, we were describing something, but this is another way of using "analogous".

1. The material from M in this paragraph has been slightly altered to fit it in—not just with the rest of this paragraph but also with the contrast made in the next.

(1) We describe a particular pattern, say, on wallpaper, by saying, "It is analogous to so-and-so."

(2) "This is the analogous case, not that."—This is quite different. For in this case we have two things before us; but in the former case we had only one thing before us and described another thing (or ordered him to do another thing) by means of the word "analogous".

The former case is like this:

2	4	6	8
12	14	16	18
102			

We tell him to continue in the same way. Suppose he writes 106 next. Then one will probably give him reasons for not writing that. One might say, "Now look, there are only 2's at the end of each figure here and only 4's here; so what are we to write here?" But he may say, "Yes, it goes 2, 12, 102 here; so it goes 4, 14, 106 here."

What do we do? At first we taught him to work according to a certain pattern. We train him in the use of the word "analogous"; we ask him to apply the training. When we told him to continue in the same way, we expected him to write certain things. So in this case our saying to him "Continue in the same way" or "Work according to the pattern" means "Write 104". And similarly "He continued in the same way" or "He worked according to the pattern" means "He wrote 104".

But in the other case, where we have the two things before us, it is quite different. When I say, "Surely writing down 104 is doing the analogous thing", "Surely this is the same pattern"—I am telling him what I mean by the word "analogous".

In the one case, if I say, "He worked according to this pattern, and put 104", I am saying something which may be false. Such a statement can be contradicted by saying that he did not write 104, that he did not work according to the pattern.

If I said, "The man who wrote 104 worked according to the pattern, and not the man who wrote 106"—couldn't I be contradicted?—"It depends on what you call working according to the pattern, applying the pattern, doing the analogous thing."—The answer to "Surely he did the analogous thing" is "It depends".

One can put the difference more clearly in this way. Suppose you are trained to use the word "analogous" to report to me what is on so-and-so's wallpaper. And I am trained to reproduce the pattern on hearing what you say. For instance, you say to me, "I saw on Watson's wall 59 multiplied by 61, analogously to those multiplication sums you showed me." And I then draw

$$\begin{array}{r} 59 \\ \underline{61} \\ 59 \\ \underline{354} \\ 3599 \end{array}$$

Here the use of the word "analogous" is to describe something, to give information about something. That is one language-game.

But now we have quite a different language-game. I point to two things in turn and say to you, "Surely this is analogous to this." The difference now is that we point to two things instead of to one. Hence this game is not to describe what is here or what is there; for we have both things in front of us and can see them.

Similarly when I say to the child, "106 is not analogous" or "Surely 106 is not analogous". I am training him to use the word "analogous".

It tells you something about "analogous". Does it tell me something about 106? What would this mean?

You might say, "You're talking nonsense, Wittgenstein; you don't know how the word 'analogous' is used. It's used for conveying information."—But why did I—quite automatically—put in the word "surely" when I said "But surely 106 is not analogous"?

Compare Professor Moore on "see"—"Surely I see in the same sense . . ." What does this mean?

Obviously this is a way of buttonholing him, trying to make him do something.

It is to show him how in this case I use the word "analogous" —otherwise I can contradict him. I have given him something like a definition. I try to give him an idea of how I'm going to use "analogous". It is part of a skill.

Return to the straight line case. He says, "I can draw a straight line through any three points, see!" and draws this:

We would say, "This is not analogous to  ——·——." What are we saying when we say that? How is that proposition used?

You might say, "In a sense it is, in a sense it isn't." If you tell me it isn't analogous, you're saying something about the word "analogous". You say what it is you are going to call the analogous step.

Instead of saying that one cannot draw a straight line through any three points, one might say, "There is no construction for these three points . ˙ . analogous to ——·——." Similarly, in saying that one cannot construct the heptagon, one is saying that in this case there is no analogue to the construction of the pentagon. In each case, one is giving the use of "analogous".

You construct the series of constructible polygons. "This construction is analogous to this."

Lewy: It comes to saying, "We're going to call this analogous to that, not to that." Then where is its importance? What's the use of making this distinction?

Wittgenstein: It seems trivial; but it is important. For we are teaching a technique of making certain kinds of patterns: pentagons, heptacaidecagons, etc. We're separating things which are entirely different, drawing an important distinction between these polygons and other polygons.—Why do I say this? (1) There is the experiential fact that we cannot easily produce regular heptagons; but that is of little importance. (2) The real point is that we cannot get a technique for constructing them.—At least, one cannot get a technique in the mathematical sense. For one may get a technique of judging what a seventh part of a circle looks like, but one cannot get what a mathematician calls a technique of construction: say, "Take three lines, divide them in such-and-such a way, etc., etc."

There are the strongest reasons for keeping apart the polygons for which one can get such a technique and those for which one cannot.—What are these reasons? Why is it important to say "This is analogous to that"?

It is important because it classifies things. This is opposed to applying a certain means, such as a jemmy, to get a certain end, such as the opening of a door.

One can look at it experientially, and then there is a difference between means and ends. In this case the means are the ruler and compasses; we can usually get a pentagon which is found to be equal when measured, and we cannot usually get a heptagon.[2] —But if one looks at it mathematically there is no difference between the means and the end. We teach him a series: the series of constructions of regular polygons. In teaching him this, we teach him a certain technique.

If we say, "You can't construct a heptagon"—you might ask, "*What* can't you do?" This presupposes that we have taught him what 'constructing' means; and by saying, "You can't construct a heptagon" we say, "There is no analogue to these constructions in the case of the seven-sided figure." We are explaining the use of the word "analogue". We could if we liked say that there *is* an analogue. But that is not what we do in fact say.

Turning: [It might be said,] "This is not the information you're trying to convey by saying one cannot construct a heptagon."

Wittgenstein: Well, in a sense it is and in a sense it is not. What is the information one wants to convey?

You might say, "No, I'm not trying to convey this information. Mathematicians don't even use the word 'analogous'."—But does one not prove that there is no analogue in the case of the heptagon?

A proof goes in fact step by step by means of analogy—by the help of a paradigm. Russell gives rules for transformations and

2. (From "One can".) This passage is based on B. But the contrast between pentagon and heptagon is from S; the wording of B has been altered so that the contrast could be fitted in.

then makes transformations. Similarly with all proofs: you're leading a man step by step, up to saying at each step, "Yes, this is the analogue here."

Mathematical conviction might be put in the form, 'I recognize this as analogous to that'. But here "recognize" is used not as in "I recognize him as Lewy" but as in "I recognize him as superior to myself". He indicates his acceptance of a convention.—When I say, "Yes, I see. I recognize that there is no construction of the heptagon", I am saying "Yes, I will accept this now; I'm going to do this", or "Yes, I see that it's the most natural thing to say, that the heptagon can't be constructed."

What does a man do when he proves so-and-so? Well, he writes certain symbols on a piece of paper. But what does he use it for? Well, he may use it to light fires with, or to sell, or to copy other proofs from. But if he copies other proofs from it, for all we have been told so far it may be useful only for wallpapering and fire lighting. The point is what it is used for in practice. Well, very often it teaches us the most useful form of expression.

Suppose I said, "Professor So-and-so found the series of constructible polygons. He found that the heptagon could not be constructed and that the heptacaidecagon could be constructed."—This is rather queer. In what sense did he 'find' something?

In a sense he did. He found, or showed, that it's no use trying. But in this sense it is not mathematical.

What did he find? What did he look for? He could have said, "I looked for something analogous to so-and-so", "I am trying to find something analogous." We might say, "He couldn't tell us exactly."

Suppose he had looked for both the heptagon and the heptacaidecagon. [In the one case] he found in the end there wasn't anything analogous. It might seem he couldn't have looked for anything.—Suppose you say that in the end he finds the word "analogous" has no use here, and that he learns its use in the case of the heptacaidecagon. We could say "He has been led to change his use of the word 'analogous'." And that is quite true.

At first when he says "analogous", he explains, "Look, this is the construction of the pentagon, and this is the construction of the hexagon; I want to do so-and-so." But afterwards he would say, "This is the analogue in the case of the heptacaidecagon to the construction of the pentagon." He explains to us a new way of using "analogous".

He said, "I'm looking for . . . etc." What did he do? He wrote down a lot of equations of a certain characteristic kind—a lot of lines and symbols.—See what a man does and then you'll see what trying to do it is.

What is the relation between trying to solve it and solving it? What happened when he solved his problem?—Suppose I said, "He has found a new kind of analogy"—an analogy between constructible polygons. How would it be intelligible to say, "He looked for it and then found it"? Isn't it absurd to say that?

Turing: It is not at all absurd. It is like "He looked for a white lion" or "a white animal between a lion and a horse".

Wittgenstein: But it is not like that. The very point of this discussion is to see the great difference.

Where is the difference? You say it.

Have you found a white animal etc. if you've drawn it? Could I draw the construction of the heptagon before I find it?

Turing: One could explain how to recognize the construction of the heptagon.

Wittgenstein: Yes, but that is very different from the description of a white lion. In the case of the white lion you can say what it will be like when you've found it. But not so in the case of the heptagon. In the case of the heptagon, it would be like describing the East Pole. The result of one's search for the construction is that one finds that the question is meaningless.

In the case of the white lion I show you what I'm looking for by analogy: "I'm looking for something analogous to this", showing you a picture of a white lion, etc. In the case of the heptagon, I give you the construction of the pentagon and say, "I'm looking for something analogous to this."

Isn't it queer—you look for something by drawing things. What the hell? You're not looking for something.

Now suppose:

"I've found the analogue to this picture of a lion."

"I've found the analogue to this picture"—showing a pentagon.

The difference is in the use of "analogous" when we describe two expeditions: a mathematical expedition and an expedition to the North Pole.

If I bring the lion into the room and say, "See, this is analogous to the picture"—I say "I have found it." If I draw the construction of the heptacaidecagon and say, "See, this is analogous to this" —what is it I've found? This figure? Was the point to bring along this object?

In fact, what had I done in proving that it was analogous? I didn't bring in the drawing, but I showed it was analogous. I didn't show that the white lion was analogous to the picture. The purpose of the expedition to the North Pole was not to show it was analogous. But the whole point of the mathematical expedition was to show that these two figures are analogous.

I've found a unicorn, not that a unicorn is analogous to this.

Construction of pentagon

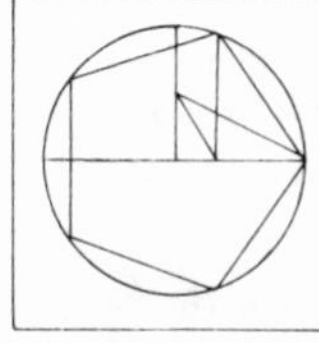

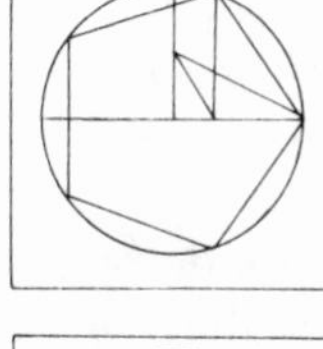

Construction of heptacaidecagon

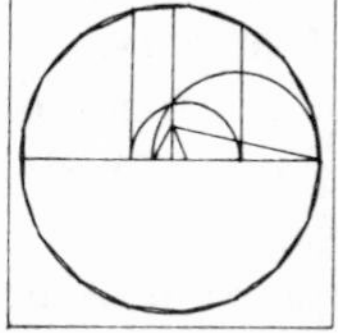

"I will call this the analogue."

"Here is the unicorn. This is the analogue."[3]

3. The pictures have been supplied by the editor. The constructions are based on H. W. Richmond's, given in H. S. M. Coxeter, *Introduction to Geometry* (New York, 1961).

What would the finding of the construction of the heptacaidecagon be in the sense in which I find the unicorn? It would consist in finding a piece of paper which had been lost and on which it was written.

"Finding the unicorn in the sense in which we find the construction of the heptacaidecagon would consist in modelling a unicorn according to the picture." The picture of the unicorn is used to model something after it and so is the picture of the pentagon. But the point is that the picture is in each case used in a completely different way.

We could put the contrast between the two cases more clearly as follows. Let's suppose we have two drawings of the pentagon—and find an analogy in each case. But we use them in different ways:

(1) To draw a picture of a pentagon from the picture.

(2) To draw the construction of a heptacaidecagon from the construction of the pentagon.—Turing might say this is drawing something in a different projection. Is this the case of the man who invented the construction of the heptacaidecagon? Does he follow a rule for projecting it (like drawing on a different scale)?

Wasn't he *introducing* a new mode of projection? He *invented* a new mode of projection, which there is reason to call so-and-so. He discovered a new kind of analogy.

He had learnt one mode of projection in the one case, but not in the other. He was given a picture. And the point was to invent a mode of projection.

Turing: It certainly isn't a question of inventing what the word "analogous" means; for we all know what "analogous" means.

Wittgenstein: Yes, certainly, it's not a question merely of inventing what it is to mean. For if that were the problem, we could settle it much easier by making "analogous" mean "cushion".

The point is indeed to give a new meaning to the word "analogous". But it is not merely that; for one is responsible to certain things. The new meaning must be such that we who have had a certain training will find it useful in certain ways.

It is like the case of definitions. Is a definition purely verbal or not?—Definitions do not merely give new meanings to words. We do not accept some definitions; some are uninteresting, some will be entirely muddling, others very useful, etc. And the same with analogy.

Is it essential that the man who invented the construction of the heptacaidecagon should have made this construction himself, that he should have drawn anything? No; he might have found the whole construction and proof written out on a piece of paper by a child of six.

If Professor Hardy found the proof on a wall, [the wall-decorators] wouldn't be the mathematicians, but he would.—What has he found, as it was all there?

Turing: He sees the analogy with the construction of the pentagon.

Wittgenstein: Yes; but what does that mean? What does 'seeing the analogy' consist of? Could there be any such thing in the case of the white lion?

If the lion had always been in the room it couldn't have been found. Suppose everyone had seen the white lion but hadn't realized it was a white lion. He suddenly realizes that this is the picture of that. But what does it come to, to say that he suddenly realizes this? He gives "white lion" a new meaning.

Turing [asked whether he understood]: I understand but I don't agree that it is simply a question of giving new meanings to words.

Wittgenstein: Turing doesn't object to anything I say. He agrees with every word. He objects to the idea he thinks underlies it. He thinks we're undermining mathematics, introducing Bolshevism into mathematics. But not at all.

We are not despising the mathematicians; we are only drawing a most important distinction—between discovering something and inventing something. But mathematicians make most important discoveries.

Unfortunately Turing will be away from the next lecture, and therefore that lecture will have to be somewhat parenthetical. For it is no good my getting the rest to agree to something that

Turing would not agree to. Hence we shall have to continue this subject in the next lecture but one.

[During this lecture Wittgenstein referred to his slogan, "Don't treat your commonsense like an umbrella. When you come into a room to philosophize, don't leave it outside but bring it in with you."]

VII

We asked: What's the difference between finding a white lion, corresponding to a picture of a white lion, and finding the construction of a heptagon, corresponding to the construction of a pentagon?

You might say, "What you're trying to find if you try to find the construction of a heptagon is a *proof*." What is a proof? Roughly speaking, you can say finding a proof is constructing a sentence or proposition—it does not matter which you say—by operating on certain given propositions, called primitive propositions, according to certain rules.—But: (1) not every construction of a sentence is a proof. You can construct the sentence "It is pitch dark in this room" according to certain rules, but this would not be constructing a proof. (2) Not every proof proceeds from primitive propositions—for example, my constructing tautologies. The idea was to give a proof not proceeding from primitive propositions.

But is a proof just constructing a proposition? Doesn't it show also that the proposition is true? But this isn't satisfactory. To say proposition p is true is just the same as to say p.

You might say, "Can't we explain what we mean by 'is true'? For example, to say that p is true means that it corresponds with reality, or that it is in accordance with reality."

Saying this need not be futile at all.—"What is a good photograph?" "One which resembles a man." We explain the words "good photograph" by means of "resemble", etc. This is all right if we know what "resemble" means. But if the technique of comparing the picture with reality hasn't been laid down, if the use of "resembles" isn't clear, then saying this is no use. For there may be many different techniques of comparison and many different kinds of resemblance. For instance, one thing may be said to resemble another if it is a projection of it; but there are many different modes of projection—of representing an object.

We may originally have a certain technique for finding whether a photo resembles a man or not. And we may extend this technique. In that case there may be many different techniques, any of which we might decide to call the continuation of the old technique.

"Always follow the old road." You can't say what is following the old road.—The order is taken from this case:

Sometimes what is meant by agreement with reality is quite clear. But in a certain number of cases it doesn't determine what we are to do.

Collating the people in this room.—I may have a list, and I may look at each person in turn and tick off his name on the list. "So-and-so, so-and-so . . . The following people are in this room." Or "The following people are sitting, the following standing," with a picture of standing and sitting, etc. This is the kind of case from which we get our picture.

But there are cases where we don't collate—for example, the tautologies of logic. And of course as the situation gets more and more complicated, God knows what process we should call collation.

How do we collate Darwin's theory? Just look. It is surprising.

The point is this. We say that some propositions are true and some false. Or, what is the same thing, we assert some and assert the negative of others, deny others. Asserting and denying are like nodding and shaking the head. And we nod and shake in all sorts of circumstances. We nod approval, nod when a dog does what we want it to do, nod agreement when someone says, "It's raining", etc., etc.

Similarly, even when we say that we assert a proposition, there are any amount of things that one can do with assertion. For instance, think what one does when one says, "Well, well, here I am." Or when the rules of a game are expressed by assertions, "One does so-and-so, and then so-and-so, etc. . . ."—Is this true or false? In the sense that it is how people play the game, it may be true or false. If you were asked of things on a slip of paper, "Is it true or false?" you'd say, "It's a rule" or "It's neither true nor false".

One might almost say, although it is not quite right, that "Yes" signified approval and "No", disapproval. "Yes"—"This is said"; "No"—"This isn't said". But why it isn't said is quite a different question.

If someone has written down a proof, constructed a proposition from other propositions according to certain rules—this doesn't tell us anything at all about the proposition he is said to have proved or about its use. What he has done might not be like a proof at all.

You might say, "It proves first of all provability" or "A proof is certainly one thing, it is a proof in the geometry of the symbolic system." It proves that from certain arrangements of symbols other arrangements can be obtained, just as the construction of the heptacaidecagon proves that the heptacaidecagon can be constructed. This is regarding all proofs as geometrical figures. It is using a proof in order to draw wallpapers.

But what is this proof that a certain proposition is provable or constructible? What is the use of it?—It might be used for many different things. For instance, it might be used to predict what the wallpaper apprentices will write.

Here is a proof:

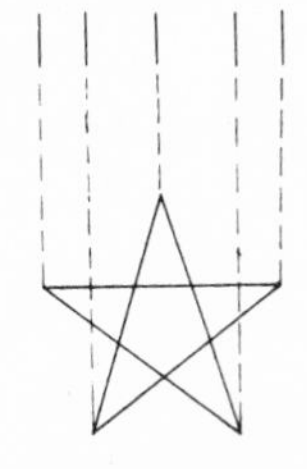

I have proved that the hand has as many strokes as the pentagram has points.

Now suppose I have two sacks of potatoes, and I fix strings so that one end of each string is attached to a potato in one sack, and the other end to a potato in the other sack, and no potato is attached to more than one string. Is that a proof?

Malcolm: I would be more hesitant about calling it a proof than about calling the other a proof.

Wittgenstein: Let's consider why.

In the former case what have we proved? That *this* hand has the same number of strokes as *this* pentagram has points? Or that *every* hand has the same number of strokes as *every* pentagram has points? The former should sound the more modest statement, because it has the narrower range. Yet in some queer way it does not sound more modest; it sounds rather strange.

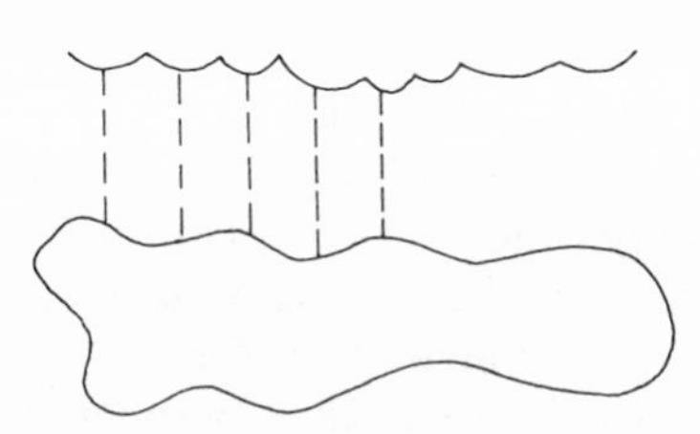

Is this a proof that *these* are equal in number to *these?* This would be an experiment; and the other figure might also be an experiment to show that one set of things is equal to another. But it can also be a proof. Now why?

It is a proof when used in a particular way.

What way?—Suppose there is a pentagram on the wall and a

hand on the floor. I say, "This is a pentagram and this is a hand; there is the same number of dashes and points." The point is that I have not reached that conclusion by correlating them; I have reached it simply by looking at this figure. One might say that we have here a new way of establishing numerical equality.

Compare "There are the same number of dashes here as there are points there" with "The hand has the same number of dashes as the pentagram has points." The former is temporal, the latter timeless. How has time vanished in this case?

Could we find one day a hand not having the same number of points? One is inclined to say, "No, for then it would not be a hand." But it is conceivable that we should in the future find that (as we should then say) we had always been drawing the lines wrong. We might have always drawn two lines to the same point by a slip. We should then say, "I must have been blind" or "I was bewitched" or something of that sort. To a man something could seem a hand, and he'd say, "Now I'm correlating one to one—and it doesn't work." Of course it is not at all plausible in this particular case; but it would be more plausible if we had a much more complicated figure.—The point is that we can imagine people in the future drawing hands and pentagrams and getting into difficulties when they try to correlate the strokes and points.

When we proved that the hand has as many strokes as the pentagram has points, we did not do the same as we do when we perform an experiment—such as the experiment with the potatoes. One might say that this figure is not an experiment but the picture of an experiment. A picture or film of an ordinary experiment is not the same as an experiment; for the film may be faked. But it [can be] a proof.[1] You might say that the relation between a proof and an experiment is that the proof is a picture of the experiment, and is as good as the experiment.—This is very important, as may be seen as follows.

1. S has "But it is a proof". But cf. next paragraph.

Suppose that I film a certain experiment. Then I may use this film as part of an historical sentence: "Malcolm did so-and-so." But I could also use it in another way. For I could say that I am going to describe all future experiments by saying that they either agree with this experiment or disagree with it by so much. It now serves as a standard; this use makes it aloof and non-temporal.—I simply say, "Lewy made the experiment, and such-and-such was the result."

Compare "The hand has the same number of strokes as the pentagram has points" and "This sack has the same number of potatoes as that sack." One can say that one describes an internal relation between the hand and the pentagram, and the other describes an external relation between sacks of potatoes.

An internal relation, one might say, lies in the essence of things. An internal relation is never a relation between two objects, but you might call it a relation between two concepts. And a sentence asserting an internal relation between two objects, such as a mathematical sentence, is not describing objects but constructing concepts.

We may say: We *accept* this figure as a proof that the hand and the pentagram have the same number. This means that we accept a new way of finding out that two things have the same number. We don't coordinate things one with the other now; we just look at this figure. I have now changed the meaning of the phrase "having the same number"—because I now accept an entirely new criterion for it.

If it should turn out that someone says he has drawn a pentagram [. . .], I will say, "It wasn't a hand" or "It wasn't a pentagram", etc.

But let's go back to the point where I said that a proof is the construction of a proposition, but that the proof does not tell us at all what is to be done with this proposition which is called 'proved'. You might say, "What a proof really proves is the compatibility of the proposition with the propositions from which one started, the primitive propositions, or rather the incompatibility of the opposite."

Russell said that mathematical propositions are of the form "If so-and-so, then so-and-so".[2] He might have said what would come to much the same, namely that mathematics says only that if the primitive propositions—which we accept as self-evident—are true, then the theorems are true. But it is not a question of self-evidence at all; it is not a psychological matter which leads us to accept certain primitive propositions. For instance, if someone said that it was self-evident to him that it never rained at the North Pole, we should not be inclined to put that among the primitive propositions of logic.

But I wanted to discuss the point about a proof proving that one proposition was compatible or incompatible with others. This is connected with last lecture's business about "analogous". For there are two uses of "incompatible" just as there are two uses of "analogous".

We can use the word "incompatible" in this way: I write down a proposition on the blackboard and say, "Lewy, write down a proposition incompatible with this" or "Lewy wrote down a proposition incompatible with this."

Then there are mathematical propositions. I write down another proposition on the blackboard and say, "This is incompatible with that."—This is timeless.

The two former are descriptive uses of "incompatible"; the latter is not.

What do I do when I prove that a proposition is the only one that is compatible with such-and-such other propositions? Well, here the same question crops up as we had about continuing a series. Suppose I write

2 4 6 8 10

and then say, "What next step is compatible with this?"—obviously, 12. And the preceding series might be called a proof that the next term is 12.—This might be part of a psychological ex-

2. *Principles of Mathematics,* §5.

periment to show what Watson calls 'compatible'; but that is not how it is used in mathematics.

We might say that in doing this we are building a road. I may tell Lewy to build a road in order to see how he builds roads or in order that we may afterwards travel along it. But in the case of continuing the series, it should be natural for all the rest of us to say "12"; we want everybody to build it in the same way.

If we say we've proved that a proposition is compatible with primitive propositions, one might say, "Compatibility is all sorts of things. We have all sorts of modes of making compatibility; we go all sorts of roads that come natural to us." [3] The question might arise, "Compatible in what way?"

$$\begin{array}{r} 25 \\ 25 \\ \hline 125 \\ 50 \\ \hline 625 \end{array}$$

Suppose I say to you, "This multiplication gives the result 625." But where does 'this multiplication' stop?—This multiplication is a pattern; and if it does not include 625 it is either incomplete or incorrect. Thus the multiplication of 25 by 25 is not a means to 625, it *contains* 625; 625 is a part of the pattern.

If you say that the proof proves these two things are incompatible—in what way?—You show me the proof. It doesn't show me they are incompatible—but that they are incompatible in *this* way. The whole pattern is a picture of incompatibility.

If I had only the beginning and end written down: "These—'25 × 25' and '624'—are incompatible." "In what way?" "In *this* way"—then I have to show you the whole sum. —This chunk isn't incompatible with this chunk. The whole form you could call a form of incompatibility.

One cannot describe an internal relation without giving both ends of the relation.

3. The passage is from S; the words rendered as "making compatibility" are "making comp" there. The corresponding passage in B has "being compatible".

We compared looking for a white lion with looking for the construction of the heptacaidecagon. And in order to eliminate inessential differences we reduced it to using the picture of the pentagon in two different ways. In the one case we used it in order to find the construction of the heptacaidecagon, and in the other case we used it in order to copy it on a certain scale or with a certain projection. We saw that the difference lay in the fact that in the former case we were introducing a new projection. But it seemed at first as if the difference lay wholly in the fact that the man who told us to look for the construction of the heptacaidecagon did not give us a very clear idea of the projection which he wanted. It was this which made Turing at first say, not "looking for a white lion" but "looking for a white animal between a horse and a lion".

Watson: Doesn't the time element enter into the case of projecting the figure of the pentagon in such-and-such a way, but not into the case of constructing the heptacaidecagon?

Wittgenstein: Yes. But time could enter into the case of the construction of the heptacaidecagon, and then it would sound queer. "Professor Hardy found the proof and an hour later wiped it off the blackboard." It sounds as if he had destroyed the proof, which is what he had found. But what he found was a technique.

Whereas in the case of projecting the figure of the pentagon in a certain given way, what we are interested in is his drawing this figure. There is nothing queer in saying he first drew the projection and then wiped it out; this would be like saying that he first found the white lion and then cut it up.

If I tell someone to draw for the heptacaidecagon a figure analogous to this figure for the pentagon, then I cannot make my order more specific. For if I tell him exactly what projection I want, then I shall have told him what the construction of the heptacaidecagon is; and if I do that, we should not say that he had found the construction of the heptacaidecagon. That was why I said that the man who discovered the construction of the heptacaidecagon had changed the meaning of the words "construction" and "analogous".

VIII

The difference between finding a white lion and finding the construction of a heptacaidecagon: some of the differences are unimportant and we tried to eliminate these by having two pictures.

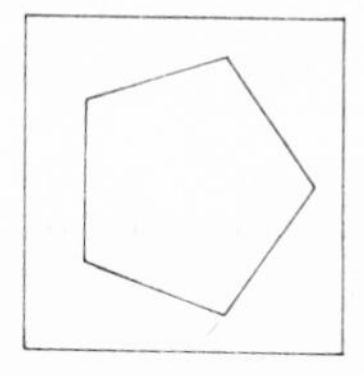

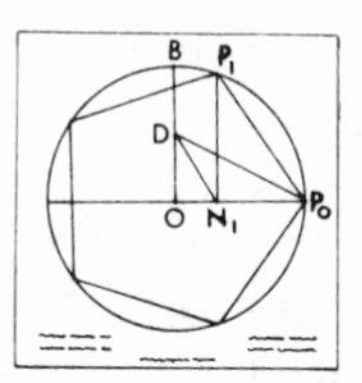

On the left a pentagon, on the right the construction of the pentagon together with its proof.[1] The problem in both cases is to produce an analogue.

(1) To produce on this wall a picture of this pentagon, say twice the size, or leaving a certain margin of indeterminacy in the method of projection; say it must be an orthogonal projection with the angle between 60° and 90°.

(2) To produce a picture: he finds the construction of the heptacaidecagon.

Now in both cases I could say, "The first picture is a picture of this." So it seems that Turing was right, and that there is no essential difference between the two situations.

And he is right:

(1) He draws a picture of what's here. He proceeds according to this technique and produces a certain picture here.

(2) He produces a certain picture there.

Each of the second-drawn figures corresponds to an original picture. There is no difference there.

1. The pictures have been supplied by the editor.

The difference lies elsewhere. Now what is the difference?

Let's ask: What's new in each case?—In both cases he produces a picture on the wall. But if one wipes away the picture of the pentagon one has wiped away what one has produced; if one wipes away the construction of the heptacaidecagon one hasn't wiped away what one has discovered—one hasn't wiped away the proof.

We might say that the important thing wasn't that Professor Hardy found this particular figure: he found a shape—as distinguished from finding something which has the shape. When he wiped it off it was still true he'd found a shape.

But is it any shape? No; it is a shape fulfilling certain conditions.—Yet this again seems to show that our two cases are alike. In the one case he has found or produced a figure, which has a certain shape, and in the other, a shape; but what is the catch here? Where does the difference between them lie?

The difference lies in the kind of conditions which the shapes have to fulfil.—Suppose we said, "Find the shape of the object in this room which is entirely red", or "What is the shape of the face of the oldest man in the room?" This is quite a different kind of condition from the condition we gave Professor Hardy when we asked him to find the construction of the heptacaidecagon.

You can say that the shape satisfies empirical conditions. The object can stop having this shape. There is a time element in the proposition: "At present, this is the shape . . ."

Again, if we say he'd found the shape which satisfies this condition—this might be false, if, say, the completely red object had another shape. That is, fulfilling this condition is an external property (relation) of the shape. But the case of finding the construction of the heptacaidecagon, of finding a shape analogous to the construction of the pentagon, is quite different. Here the shape has an internal relation to the conditions. We may ask, "What would it be like if Turing found a shape which fulfilled these conditions but which wasn't this shape?" In the one case, if it weren't like this, it wouldn't be the construction, while in the other case it would be easy to imagine such a shape.

But one must not think it is as if between Turing and Watson there are both internal and external relations, and that we are to ask which of the relations which hold between them are internal and which are external.

I look through a telescope, take positions, do calculations, get a result, look again—and the star is still there. What I've discovered is a technique. Suppose someone suggests splitting the discovery up into a mathematics part and a physics part.

A man wants to lay a floor in a room. He wants to know how many boards to buy. He takes a tape measure—which he found on a tree—and stretches it along the room and finds a certain number at the end. He stretches it across a board and finds a certain number. He discovers *dividing*—dividing the first number by the second. Then he says to the timber merchant he wants the number of boards which he gets by this operation—and then makes a forecast that they will fit; and they do.

He's made a discovery. He did certain things with tape measures, then with numbers. What is the discovery?—He has certainly made a discovery in physics. There is a temptation to say, "He made a discovery in physics and besides this a discovery in mathematics." But let us pick out the mathematical part of the discovery from the physical part. Should we say, for instance, that boards do not vanish into thin air and also that 125 ÷ 5 is 25?

He developed a calculus for this purpose, but knew no other part of arithmetic. Has he discovered that 125 ÷ 5 is 25? This is very queer. For we know what it would be like for him to discover that boards do vanish into thin air; but what would it be like to discover that 125 ÷ 5 is 19?

Couldn't he have calculated that it wasn't 25 but 19? What justified him in saying that 125 ÷ 5 is 25 was that it gave the right result. He could just as well have under other circumstances 125 ÷ 5 is 16.—He has produced a technique for making patterns; and '25' is part of this pattern. One cannot say that he has discovered that this technique gives this result; for the result is part of the technique. One can imagine a pattern in which all the rest was the same but the part we call the answer or result was different; this would be a different technique.

We would feel inclined to say that it isn't dividing. But that doesn't get us anywhere. For it makes it look as if we already had an idea of division and then applied it to this case. But what is the criterion of dividing—how do we define 'division'? If we don't define it as a certain technique ("As ordinarily used, $125 \div 5$ is by definition 25")—then we shall have to define it by saying that it is the process which gives the right number of boards. But that would make it empirical, which it is not.

Multiplication could be defined by an empirical criterion. If you have 16 rows of soldiers, 19 in each row, the result by multiplication will be the same as by adding.—One feels inclined to say that if he reaches a different result from such-and-such, then he cannot mean the same by the signs as we ordinarily mean by them. "If '×' means the same, then 16×19 must have this result." [2]

Take an example from logic:

Suppose we say that $\sim\sim p = p$; double negation equals affirmation. But why should we not say that $\sim\sim p = \sim p$? For double negation is sometimes equivalent to negation.—But then we are inclined to say, we can't have both. If one person uses it one way and another in another way, then it must mean something different in the two cases. (This is what we said about the use of the division and multiplication signs.)

But must it?—This sort of talk comes from the case where I pour something from two bottles in turn on two pieces of zinc and they give different reactions, from which I conclude that the bottles must have contained something different.

Or at any rate one is inclined to say that either '$\sim$' must mean something different in the two cases or else the signs must be combined in different ways. For instance, one might explain:

(1) $\sim(\sim p) = p$

(2) $(\sim\sim)p = \sim p$.

But who says how we're going to use brackets?—This comes from thinking that the meaning of a sentence is a complex which is composed of simples and their combinations, just as a table is composed of its various parts arranged in a certain way.

2. Most of this paragraph is only in S and is in part questionable. Material from B, which probably referred to the same example, has been included.

We are inclined to say that '∼' must mean something different in the two cases—and there is some truth in it. For we might explain that double negation equals affirmation by a picture of turning something 180° and then another 180°:  . And we might explain that double negation is sometimes negation by a picture of putting the same thing down on the mantlepiece twice:  .[3]

The pictures which come to mind in the two cases are different. Is this unimportant? No. But does it get us any further? No. For these pictures are only another set of symbols.

We could use these pictures:  and  , instead of negation signs, and we could write $\longrightarrow p$ for p. We might thus show a man clearly that double negation often is affirmation: that negation consists in reversing the direction of the arrow. But that is still a symbol; I have replaced one symbol by another.

If the *meaning* of a symbol is something like a picture suggested by it, then one can say that '$\sim p$' means something different in the two cases. But this is not conclusive for many reasons. (1) There may be no picture at all suggested by a symbol. (2) People may in an overwhelming majority of cases react in one way to a certain picture and yet it may be possible for it to be used differently. For instance, pointing normally suggests to one to go in the direction from the shoulder to the hand, but it might suggest to one to go in the opposite direction. Similarly, a railing round something generally suggests to one an invitation to keep out and not an invitation to jump in, although when one is on horseback a railing in the middle of a field may suggest to one to jump it.

Is the picture a sufficient criterion? If I say to a man, "Get out" and he imagines himself coming in, but does go out, what does he mean by the symbol? That is, the *use* of these symbols is a criterion of their meaning, and the question is how do these two criteria work together?

If the meaning is represented by the use of the symbol, it is no use saying, "The use is different, *therefore* the meaning is different." In some cases we can even say that the meaning is not

3. The diagram has been supplied by the editor. See below, Lecture XVIII. Cf. also *Bemerkungen über die Grundlagen der Mathematik* (Frankfurt, 1974), p. 102.

different, in cases like $\sim\sim p = p$ and $\sim\sim p = \sim p$, for instance. In that particular case it would be very queer to say that negation was a different thing in the two cases.

In the case of the discovery about the planks, the man had constructed a calculus. He did this and he got the right result—that is, physically right, the planks fit the floor. But if he teaches the calculus to his child, the right result won't be the one which gives the physical result—but will be *this* result, the one which he, the father, gets. 'The right result' in this latter case means a certain definite figure.

In mathematics a description (in Russell's sense) means the same as a proper name. 'The number which is got by multiplying 5 by itself' means the same as '25'. 'The number of my shoes is the number one gets by squaring 5' is really the same as 'The number of my shoes is 25'.—By saying that they mean the same, I mean that I could substitute one for the other in any ordinary experiential sentence—but not in any mathematical sentence. The mathematical sentence '$5^2 = 25$' gives a rule that in experiential sentences you can put '5^2' instead of '25'.

Let us return to the discovery. The point was whether we should say that part of the discovery was a mathematical discovery.

It was not a discovery that $125 \div 5 = 25$; for this result is merely part of the use of the symbols.—This has to do with what I said, that 'mathematical discoveries' are better called inventions. He invented a technique; the reason why the technique is interesting and useful is an extra-mathematical consideration.

Remember the business about continuing a series—for example, the series of cardinal numbers. Here 'intuition' is the word that corresponds to 'discovery'. People say I know by intuition that 13 comes after 12.

Suppose someone says that 15 comes after 12. We would say, "That isn't the series of cardinal numbers." But then doesn't this amount to saying that part of the definition of 'cardinal number series' is that 13 comes after 12?

You might say, "13 comes in one series, 15 in another." But isn't this what makes it a different series?—In one technique one follows, in the other, the other follows. [There would be] no discovery that 15 follows 12; it would just be a technique.

Inventing a technique:

12, 14, 13, . . .

This would be immensely impractical, inconvenient—but not wrong.

Suppose I always left out 13 in my mathematics. You might say—(a) that it's useless; (b) that it's uninteresting. And under normal circumstances it would be. But if there were people who were terrified of the number 13 this mathematics might be of great importance to them. When counting nuts, people would say that one of the nuts had disappeared—that the devil had taken it away, or something of the sort. And some theological arguments are in fact of this form.

It may under certain circumstances be very useful to *count* differently—if, for example, things do disappear regularly in certain ways. In that case one is adapting one's technique of counting to the circumstances.

Again, it would normally be considered detrimental to measurement if the measuring rod expands; but we can imagine circumstances in which it might be useful. For instance, if one had to take some furniture from one room to another, the two rooms being permanently at different temperatures, and the furniture of easily expansible material—then it might be useful to have the measuring rod of the same material as the furniture.

Again, it may be very useful to me to have a measuring rod which I can pull out in order that I can cheat you when selling you something. And we can imagine a society in which that is not considered cheating and that it is thought right that the strongest grocers, who can pull the measuring rods out furthest, should do the best business.

There is no discovery that 13 follows 12. That's our technique—we *fix,* we teach, our technique that way. If there is a discovery—it is that this is a valuable thing to do.

One might be asked to project a figure according to a certain projection, and one may not know what the projection will look like. Then one draws it and finds that it is a conic of a certain sort. The difference between this and finding the construction of the heptacaidecagon is that the latter is like a riddle in a way the former is not. For in riddles one has no exact way of working out a solution. One can only say, "I shall know a good solution if I see it."

Suppose I say, "Multiply 26 by 89" or "Project the circle in a certain way." I may be giving you a mathematical task or a non-mathematical task. If it is a mathematical task, you can go away and do it elsewhere; you can do this technique anywhere. And the result is that so-and-so times so-and-so is so-and-so, which is timeless and serves as a paradigm.

But we might want this as an ornament on this particular blackboard. Then the result is that a certain figure stands over there.

Suppose that one is told to project a certain figure onto a certain wall. Then one can either do this by making its shadow fall onto the wall in a certain way or else by working out (as one might say) what would happen if one did make its shadow fall thus.—But suppose one is given as a mathematical [task] the projection of the figure onto the wall, and told to do it in a certain way, for example, by means of light and shadow. What is the difference between the mathematical task and the non-mathematical task?

We can teach a person to multiply and then say to him, "Multiply 19 by 365." Or one might show him one multiplication sum and say, "Now do the same for 19 and 365", in which case he has to invent for himself the technique of multiplication.

Similarly I may draw the construction of the pentagon on the blackboard, and then say to someone, "This is the construction for 5; now do the same, the analogous thing, for 17." If he had learnt the series of constructions of regular polygons, my order would have referred to a certain technique which he has learnt. But if he has not learnt it, he may invent a technique for this.

There are *many* things he might do, many analogous things. There might even be a proof that nothing analogous *could* be done for 17.

So what is done depends on the meaning of 'analogous'. How is the meaning of 'analogous' fixed? (1) By giving a few examples other than this, leaving it to him to apply them to this case, or (2) we give *this*.—We tell him exactly what 'analogy' means in this case. Then we have given him the answer and the order makes no sense.

We may leave it open.—How is what he is allowed to do still fixed? One might say that it is by the applause he receives if he gets a certain analogy. But the problem is not to do what will please such-and-such people—that is not a mathematical problem.

Then how is what he is to do fixed? Certain things we will decline to call analogous, of others we will say they are analogous in an unimportant way, of others we will say they are analogous in an important way.

You might say, "If he was very clever, he understood us—understood what the analogy was." But you can't give an internal relation except by giving the two things between which it holds.

What he produced was a new form of analogy, preferable to others he might have produced. In what way is the new form preferable? Isn't it very like saying that saying '13' after '12' in counting is preferable to saying '15' after '12'?

The analogy he was to produce wasn't given him. In fact he produced it. You give him an explanation of the analogy—but you haven't taught him what to do unless your explanation works. He doesn't cross the bridge until he gets there.

" 'Now construct the analogous thing for 17.'—He might have done many things, might have produced many different sorts of analogies—or even proved it was impossible." If he did what we ordinarily do, we should applaud him—for very good reasons. But these reasons are not that the other things he might produce are not analogies.—We might say that the others are not analogous in this way, or that what we wanted was this analogy. But what does 'analogous in *this* way' or '*this* analogy' mean?

Just as he doesn't discover that 15 follows 14, but learns it or invents that series, so he does not discover this analogy; he invents the analogy or learns it. It's a good analogy—not because it's *this* analogy, which is nonsense—but because it's useful, etc.

IX

If you were told that Smith drew the construction of a pentagon on the wall—how would you satisfy yourself? What would you answer?

If you were told that Smith drew the construction of a heptagon on the wall—you might answer that this is certainly false.

Let us consider how this sort of impossibility is proved.

Suppose that we have a method of constructing polygons which is narrower than methods with ruler and compasses—say:

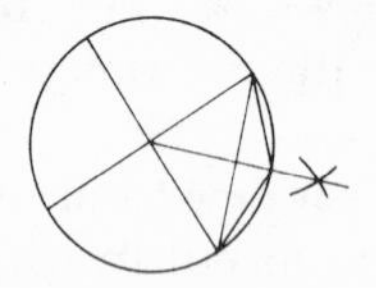

We are only allowed a ruler and a pair of compasses whose radius is fixed. We draw two diameters at right angles to one another in a circle; this gives us an inscribed square. We then draw arcs from the intersection points of the drawn diameters. Whether we call this bisecting or not doesn't matter. This is what we do. Thus we get the octagon, for instance. Similarly we could get a polygon with 16 sides, and so on.

Now someone is asked to produce the 100-gon this way. At first he goes on trying and trying, keeps on bisecting smaller and smaller angles and doesn't get any satisfactory result. Then in the end we prove to him that the 100-gon cannot be constructed in this way.

It seems as if we first of all made an experiment which showed that Smith, Jones, etc. could not construct a 100-gon in that way, and then a mathematician shows that it can't be done. We get apparently an experimental result, and then prove that it could not have been otherwise at all.

But there is something queer about this. For how could the man try to do what could not be done?

I want to ask: In what sense does the proof show that you can't do it?

The proof might be this: we go on constructing polygons and being very careful to observe certain rules. We should then find that the 100-gon is left out. If we want to construct the *n*-gon in that way, *n* has to be a power of 2. The last power of 2 before 100 is 64, after that is 128, and so 100 is left out. This would have the result of dissuading intelligent people from trying this game.

One thing which dissuades us from doing something, is making an experiment. Suppose that I try to lift a weight and find that I cannot, then I give up. That is finding by means of experiment that I cannot lift it.

Have we in some sense made an experiment and found we could not divide it?

Note that before he had the proof he tried to find the construction; after he had the proof he gave it up. If we ask what is the function of such a proof, what does it do to him—well, that is *a* function of it.

Turing: Isn't one of its functions to give me a clearer idea of the sort of thing which would happen if I constructed polygons in this way?

Wittgenstein: Yes, but what is referred to as 'the sort of thing which would happen'? For instance, he is not taught that when he tries to construct polygons in this way there is not an explosion.

The proof gives him a very much clearer idea of what he is trying to do.—Suppose I said that it changes his idea of constructing an *n*-gon by this method. What does that mean? Well, if someone asked him what he was trying to do, he would now give an entirely different sort of explanation. Before, he would have said, "Oh, I do this sort of thing" and begin to draw a few circles and lines. But now: "I am trying to see whether so-and-so is a power of 2."

If you make an idea clearer, do you *change* it? At first he could not have given the same explanation of his method which he gave later: he had a *rough* idea of it. Later he gave a different explanation—so we may say his idea changed. Or shall we say, "This was always in his head"?

The question is: Why should we call this new idea a clarification of the old one? We might say instead that later on he tried to solve a different problem or perhaps gave up the problem altogether. But we might say it's the *same* problem.—You might say he has been led to change his question. This particular method has been pointed out. After this he says, "Yes, that's what I had in mind."

Suppose we are asked, "Can the quintic equation be solved with radicals?"

I might have an idea of what a radical is, consisting just in the fact that I could give a few examples and say "and such like". But you might order these in a series, and say, "Well, it's something in this series: . . ." Then you have changed my idea.

What does it mean to change my idea in this way? When giving me the series you might ask, "Is this still what you meant?" If I say, "Yes, that's what I had in mind"—of course I hadn't had it in mind—or else you haven't taught me anything.—*BUT* I am ready now to have it in mind. I am prepared to change my idea in this way.

Again, the importance of the proof that trisection is impossible is that it changes our idea of trisection.—The idea of trisection of an angle comes in this way: that we can bisect an angle, divide into four equal parts, and so on. And this leads to the problem of trisecting an angle. You are led on here by *sentences.* You have the sentence "I bisect this angle" and you form a similar expression: "trisecting". And so you ask, "What about the sentence, 'I trisect this angle'?"

But suppose we had never tried to bisect or quadrisect, but we had immediately learned the *n*-sections of the angle as a *series,* just as we learn the series of cardinal numbers. Then the question wouldn't have arisen: we would never have to prove that trisection was impossible, any more than you would have to prove that ½ is not a cardinal number.—If we had learned from the beginning the series of constructions of *n*-gons, then nobody would ever have asked whether the heptagon is constructible. It's not one of these—that's all. And the phrase "construction of the *n*-gon" would have meaning only when *n* has 5, 17, etc. as values. —The problem arose because our idea at first was a different idea of constructing the *n*-gon, and then was *changed* by the proof.

Turing: But there would still be a problem.

Wittgenstein: Yes, that is the point. It would come in another form and not in this form. There would be different problems.

Return to the construction of the 100-gon by bisection. What was it that made him accept my proof that what he was trying to do could not be done? *Not* that something looking like a 100-gon could not be constructed—that obviously could be done.

Well, he recognized the importance of going systematically through all the angles, and not just bisecting one here and one there. Why did he recognize this?

Suppose we say: I *could* have given him a new idea altogether of constructing it. Why then should he give up the attempt to construct the 100-gon? I taught him a new method, procedure, technique. The result of this was that he constructs certain regular *n*-gons but not the regular 100-gon. And he suddenly gave up trying to construct the regular 100-gon. Why?

One can put the point more clearly thus. Suppose that I didn't try to dissuade him, didn't have him up for doing what he did, in a slapdash way, etc. I spoke as though I never thought this would change his idea at all, or affect his problem. Suppose I simply showed him my series of constructions, getting to the 64-gon, then the 128-gon, and after that he gave up trying to construct the 100-gon. Now why should he? I never said he should.

It isn't that he has tried to apply the new technique to the 100-gon and couldn't do it—for he can't try it on the 100-gon. It goes right past the 100-gon.

He saw my construction, and it no longer interested him to construct the 100-gon. He *might* have gone on trying to find approximate constructions. If he gave up, he gave up because he acknowledged *this* to be the method of construction he had always wanted to follow, but had thought of in a vague way.

What the proof does is to change his way of looking at it. It gives him something very important: this series of constructions; and he acknowledges that nothing else is what he wanted.

There are all sorts of reasons for this. For example, as a matter of experience, if he follows the method I teach him, he will get more things looking like regular polygons. But it is not merely that. Similarly, it is not merely the fact that by messing about with ruler and compasses he will hardly ever get a trisection of the angle which makes him give up trying.—There are reasons connected with the single steps of the proof and their similarity to other proofs he has made. So with the proof that the diagonal is incommensurable with the side of the square 1 × 1. The result itself is almost negligible.

[Watson brought up the case of a man who put his hands underneath his feet and pulled and said, "I am trying to lift myself."]

Wittgenstein: Well, one can imagine two cases here. Suppose that when he did this he did in fact rise into the air. He might then be satisfied and say, "I have lifted myself up" or he might not. If he were satisfied, one could only show him that this is different from other cases in which we say that we have lifted so-and-so—because, for instance, the arms do not move relative to the rest of the body. If he were not satisfied we might say to him, "What did you want? What was it that you were trying to do?" It is possible that in reply to this he would say, after thinking for a bit, "I see; I did not want anything. I was misled by an analogy." How did he get to saying that?

If someone said, "Smith constructed the heptagon" you might say, "Well, he doesn't seem to know much about it", or "He's one of those people who try to find the . . . in spite of everything." If you replied, "That's certainly wrong"—is that the ideal way of putting it? What you would think would be "There is some muddle there" or "I don't understand what you mean." This sounds very different from "It's certain what you say is false."

Compare: "Smith constructed the woohoo." We wouldn't say, "This is certainly false", but "This is nonsense."

The most transparent thing to say is: The mathematical proof has made us adopt a phraseology in which we cut out the phrase "construction of the heptagon", for very good reasons.

Suppose I get letters and put them in boxes labelled 'bills', 'love letters', etc. Then if I label a box 'honorary degrees', I do not thereby give myself any letters conferring honorary degrees. But if I go into an office and look at the labels on the drawers of a filing cabinet, it may give me some idea of the sort of correspondence and reports they get.—If we adopt a new phraseology in mathematics, this is like adopting a system of filing. Why it is done is because there are such a lot of things to go into it. It is not utterly independent of experience.

I want to go on with a different question. It often looks as though a mathematician started from a hypothesis and then later finds the proof.—We might have Fermat's problem and try lots of numbers and never find one which suits. "We have tried, and so far it hasn't worked. We may one day find a proof that this *could* never be the case." Here we seem to have something like a mathematical experiment.

Connected with this is another question. Professor Hardy has said, in an article in *Mind* called "Mathematical Proof", "Mathematical propositions state objective facts outside myself." [1] One of the arguments is that you can *believe* theorems, more or less as you believe other things. That is immensely important.

Next time I'll talk about this: How can we believe that 25 × 25 is 624, or even that it is 625?—Well, one can, of course. One can say, "What is 25 × 25? I believe it's 625; let's see", and proceed to work it out.

1. *Mind,* 38 (1929), 4.

You might say, "Isn't it queer that he can believe something false here?" Why is it queer?

Cunningham: Isn't it queer because the expression "$25 \times 25 = 624$" is meaningless?

Wittgenstein: Well, this 'meaningless' road has now been trodden so often that it has become muddy and one cannot see one's way clearly; it needs rolling.

One can ask, How deep does his belief go? How far does he believe that 25×25 is 624? How much of the multiplication sum does he believe? Does he just say, "25×25 is 624"? Or does he go on to multiply it out? And if he does multiply it out, does he do the whole sum correctly except that he writes down the bottom line as '624' instead of '625'? And if so, what does he believe that's wrong? One might say, in fact, that "He believes that $25 \times 25 = 624$" may correspond to many different states of affairs.

X

Sometimes it seems as though mathematical discoveries are made by performing what one might call a mathematical experiment. For example, the mathematician first notices a certain regularity and then proves that it *had* to be so. And this seems to point against what I said, that perhaps what we call discoveries in mathematics, would better be called inventions.

You might say, "Come, a child when he calculates 25×25 and gets 625 doesn't *invent* this. He finds it out."

Of course he doesn't invent the mathematical fact—it would be absurd to say that. And there is nothing wrong in saying that he found it out.—But the analogy which springs to mind is that of finding something by making an experiment.

Now is the child making an experiment?

Lewy: In a sense he is and in a sense he isn't.

Wittgenstein: Well, yes—but in what sense is he and what sense isn't he?

Turing: Isn't it more like an experiment when one is familiar with the rules of multiplication?

Wittgenstein: Are you making an experiment by having a coat across your knees?—Normally one would say "No". But might you not be making an experiment by it? For instance, you might be seeing how long it would be before you got too hot. And the difference between the case where you were making an experiment and the case where you were not would not lie in the way that the coat was lying across your knees. The difference would lie in the surrounding circumstances. Therefore might we not say, "Well, the multiplication might be an experiment"?

What circumstances would make the multiplication into an experiment?

Turing: One might say beforehand, "Let us see what 136 times 51 is."

Wittgenstein: Well, yes—but first let us see that it is not just a question of whether it is an experiment or not.

If we said, "Let's see what happens when we multiply 136 by 51", it may be an experiment—but it isn't clear *what* experiment. I may want to see if you can multiply correctly, or to see if the chalk will stand the strain. It may be all sorts of experiments; or going through the multiplication may be a pastime.

Suppose we don't mean these things and yet call the multiplication an experiment. Now an experiment has a *result.* So does a calculation. If one calls something the result of the calculation, is that same thing the result of the experiment?

Watson: Not necessarily.

Wittgenstein: No, not necessarily—but is it even possibly? Turing has called the calculation an experiment. But what would it mean to say that the result of the experiment is the result of the calculation?

I might say I made an experiment to see what he would write down in the end. The result of the experiment is then: that he wrote down 6936. But if this is an experiment, could you say that the experiment was wrong if he wrote down 6935? If I am merely trying to find out by experiment what he will write down, it does not matter what he writes down.

Turing: If he wrote down 6935 one would say that one ought to have arranged the experiment differently.

Wittgenstein: Well, but do you arrange the result? The experiment is to find out the result. 'A wrong result in an experiment'—what is that?

Watson: One might say in Turing's defence that one wanted to see what he would get if he obeyed certain rules.

Wittgenstein: Yes—*unless* we include in 'obeying the rules' getting just this result in this case. We can say, "We've taught him these rules; let us see what result he gets if he obeys them." But then obeying the rules must be something which may lead to the one result and may lead to the other.—We might say that we want to see how he obeys these rules.

Suppose you had made the calculation beforehand. Then you will already know what he will *have* to get if he obeys the rules. Then it will not be an experiment to see *whether* he gets so-and-so *if* he obeys the rules, but to see whether or not he obeys the rules. And one cannot then say, "This experiment has taught me that if he obeys the rules he will get this result."

I could say: Different people when given the order "Copy this line:  " will do different things: some will draw it on a large scale, some on a small scale, etc., etc. I can then make an experiment to see what Smith will do when asked to copy it.

Similarly, suppose that the phrase "solving the equation" were to mean "writing down one of the roots of the equation". Then one can, by asking you to solve a certain third degree equation, make an experiment to see *which* of the roots you will write down. But I cannot calculate which of the roots you will write down.

I cannot make the result of the experiment at the same time into the result of the calculation. If the result is the result of the calculation, I have already fixed what I call 'obeying the rules' by my calculation. The calculation gives me a form of expression now: and *now* I say he gets either the right or the wrong result. —And the result of the experiment will then be not what result he will get if he follows the rules but whether or not he will follow the rules.

Suppose we in this room are inventing arithmetic. We have a technique of counting, but there is so far no multiplication. Suppose that I now make the following experiment. I give Lewy a multiplication.—We have invented multiplication up to 100; that is, we've written down things like 81 × 63 but have never yet written down things like 123 × 489. I say to him, "You know what you've done so far. Now do the same sort of thing for these two numbers."—I assume he does what we usually do. This is an experiment—and one which we may later adopt as a calculation.

What does that mean? Well, suppose that 90 per cent do it all one way. I say, "This is now going to be the right result." The experiment was to show what the most natural way is—which way most of them go. Now everybody is taught to do it—and *now* there is a right and wrong. Before there was not.

It is like finding the best place to build a road across the moors. We may first send people across, and see which is the most natural way for them to go, and then build the road that way. Before the calculation was invented or the technique fixed, there was no right or wrong result.

When the experiment was tried on Lewy, he did just the same as a child does who is working out what 123 × 489 is. But the one was an experiment and the other was not. If you say that you make an experiment to see what result the rules will lead him to, this is only an experiment so long as the rules do not prescribe what it *has* to lead him to—so long as there is not a right and wrong. We say of the child, not "He has followed the rules in this way" but "He has followed the rules."

[Wittgenstein asked Turing a question.]

Turing: I see your point.

Wittgenstein: I have no point. If you want to interpret the word "experiment" in a wider sense, then by all means do so. And let us see whether what I have been saying may not be false.

Turing: What about the case of a man who can count and who cuts nine sticks into fifteen parts each and counts the number of parts. Then that is an experiment.

Wittgenstein: I would not contradict you on this point. But does it depend on the circumstances? Your man chops sticks and utters words; is that always an experiment?

Turing: I see that won't do.

Wittgenstein: No—but under what circumstances would we call it an experiment?

Wisdom: If I multiply 165 X 138, I should call it an experiment if I used the result to predict what other people will get if they multiply the same figures.

Wittgenstein: Yes. Now take Turing's sticks. What was it that the man who chopped up the sticks and counted has found out?

Turing: Perhaps he has found out what other people will call 9 X 15 when it is written as one number and not as a multiplication.

Wittgenstein: If he has found that out, he has performed an experiment similar to this: I want to find out what people will do if a stick is put across the door about a foot from the ground, so I ask Smythies to put a stick across the door one day when we are not expecting it. The result is that I come in and fall over it. And I say, "Now I see what happens—people stumble over it." And I may infer that it is a bad plan to put a step in the middle of a dark corridor.

But this sort of experiment is essentially different from a calculation. I can multiply 423 X 763 in order to see what other people will get—to forecast this. But then if I say, "Most people, if educated in this way, will get such-and-such", is this the result of a calculation?

Turing: One could make this comparison between an experiment in physics and a mathematical calculation: in the one case you say to a man, "Put these weights in the scale pan in such-and-such a way, and see which way the lever swings", and in the other case you say, "Take these figures, look up in such-and-such tables, etc., and see what the result is."

Wittgenstein: Yes, the two do seem very similar. But what is this similarity?

Turing: In both cases one wants to see what will happen in the end.

Wittgenstein: Does one want to see that? In the mathematical case, does one want to see what chalk mark the man makes? Surely there is something queer about this.—Does one want to see what he will get if he multiplies, or what he will get if he multiplies correctly—what the right result is?

Turing: One can never know that one has not made a mistake.

Wittgenstein: Russell said, "It is possible that we have always made a mistake in saying $12 \times 12 = 144$." But what would it be like to make a mistake? Would we not say, "This is what we do when we perform the process which we call 'multiplication'. 144 is what we call 'the right result' "?

Russell goes on to say, "So it is only probable that $12 \times 12 = 144$." But this means nothing. If we had all of us always calculated $12 \times 12 = 143$, then that would be correct—*that* would be the technique.

But let us go back to Turing's comparison of an experiment in physics with a mathematical calculation. Let us have a case of calculating by means of a balance. Suppose that one invents one's arithmetic in such a way that $2 + 2 = 4$ is proved by putting two bits and then two more bits in one scale pan and four bits in the other and seeing that neither pan goes down. In what circumstances should we call this an experiment and in what circumstances a calculation? How would 'right' and 'wrong' be introduced here?

Suppose that you put two balls and then two more balls into one pan, and four balls into the other. It is quite possible that the lever will tip over. Will you call this adding wrongly? Obviously not; it has nothing to do with adding wrongly. Or we may put three and two on the one side, and on the other side four, and the lever will be in equilibrium—and you will not then say that 3 and 2 is 4. Or are you bound to say anything that comes out?

If you are *bound* to say, "Then 2 and 3 is 4", why aren't you when a child writes $2 + 3 = 4$?

How then does right or wrong come into the experiment? You might say that you had done the experiment wrong because you had forgotten to dust the balance. But that is irrelevant because the dust will not show only in the result. You may have neglected certain things in weighing—but you need not have.

"Suppose I do this again"—here the 'this' doesn't include this result, otherwise it is not an experiment, but a calculation—there is an internal relation. The conditions of the experiment don't include the result.

But suppose I do this again and the lever tips over. Then we may say the weights have changed, or just that we don't know why it happened.

But now what happens if we make the weighing into a calculation? We should have to decide a wrong and a right result of the weighing.

It is enlightening to look on a calculation as a picture of an experiment.

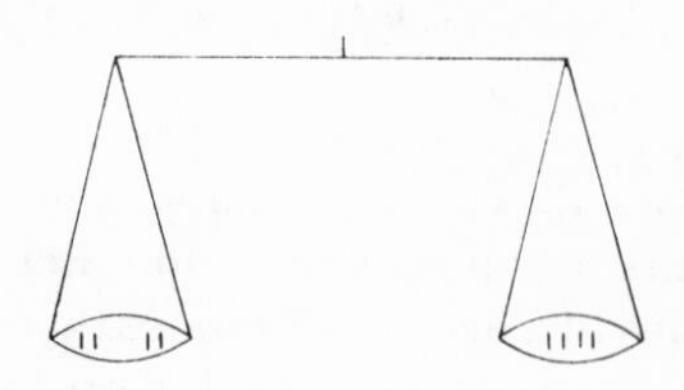

We can say that this picture is the right result, and we can say that Jones got the wrong result, meaning that he did not get this result.

If it is a calculation, we *adopt* it as a calculation—that is, we make a *rule* of it. We make the description of it the description of a *norm*—we say, "This is what we are going to compare things with." It gives us a method of describing experiments, by saying they deviate from this by so much. (Compare our previous example of the stars.) [1] If we call it a calculation, it's a complete picture which now serves as a standard or phraseology for the description of an experiment.

We might have adopted $2 + 2 = 4$ because two balls and two balls balances four. But now we adopt it, it is aloof from experiments—it is petrified.

(It may not have been either experiment *or* calculation. You did certain things, made certain noises, etc.)

Yet there are cases where we talk of a right and a wrong result to an experiment. For instance, it may be that if a pupil in a laboratory mixes H_2S and SO_2 in such-and-such proportions and does not get a bang, we say that he has not got the right result.

1. See Lecture IV.

Now suppose that we have a picture of this experiment, as one often does have pictures of experiments in scientific textbooks. Or better still, suppose that we have the chemical equation:

$$2HCl + 2Na = 2NaCl + H_2.$$

This would not ordinarily be taken as a proposition of mathematics. It would be experimentally verified.

But it could play the role of a proposition in pure mathematics—if it were in the end taken as a way of describing how the experiment had come out.—Could I make this independent of the result of experiments (that is, a mathematical proposition) —although it's true that experiments gave me the idea? If I begin to use it in such a way that I would consider this correct *whatever* the result of the experiment was, then it would now sink (or rise) into the role of a rule; and we now describe experiments by means of this rule. We'd say "In one experiment something vanished", etc.

Turing: Suppose chemists were to write a large book of equations and decree that these should be the correct ones. Then I could look in the book to see what the right result is. This will be an experiment.

Wittgenstein: To see what is in the book?

Turing: No. To see what the right result is.

Wittgenstein: Well, what is the result of this experiment?

One might make the experiment with many different people in order to see whether, for instance, they can look up equations properly—and then there is no right and no wrong result of the experiment.

You said that looking an equation up in a book is an experiment. Of course, it may be an experiment, just as anything may be an experiment. But why do you call it an experiment in this case?

Wisdom: Children in a school laboratory are doing experiments, and yet if they get a certain result we say, "You must have got it wrong."

Wittgenstein: Well, it is a good test for whether a thing is an experiment or a calculation to ask whether it is just as good if we do not do the experiment but just paint a picture of it.

There is a temptation to say that a thing is not an experiment unless no one knows the result. But that is obviously wrong. No one uses the word "experiment" thus.

We teach people to weigh. We do not teach them to get such-and-such a result. We teach them to adjust the balance beforehand in such-and-such a way, not to woggle the table, etc., etc. —Now compare "You must have woggled the table" with "You must have made a mistake in your calculation." Wisdom is right that the appearance of the phrase "must have" is not an infallible proof that we are dealing with a calculation and not an experiment.

[Wisdom brought up the example of "He is trying to find out whether one can mate with two knights, a king, and a bishop", as compared with "He is trying to find out whether all French verbs of motion take *être*."]

Wittgenstein: The former would generally be called a calculation—a picture would do as well.

Suppose he moves pieces and then says, "Therefore one can mate." This shows it is a calculation. We could put it in terms of 'temporal' and 'non-temporal'. The calculation does not give a temporal result; it does not show that one can mate now. It is a picture of mating, of what I am going to call 'mating'.

Looking in the chemist's book of equations—this could have been an experiment or anything. Normally we won't call it an experiment, because it is not to see what he will get. Is going over the moors an experiment? Not usually; but it may be, for instance if we want to see which way a certain man will take when told to go to such-and-such a place, or which is the natural way to take.

Watson: There is a temptation to say that in a multiplication the rules do not tell us what the last line is but that the last line follows necessarily from what the rules do tell us.

Wittgenstein: Yes, certainly. We have learned the rules of multiplication, but we have not learned the result of each multiplication. It is absurd to say that we invent $136 \times 51 = 6936$; we *find* that this is the result.—But the catch in what you suggested comes in with the right and the wrong result. For when I multiply, do I want to find out what result I shall get or what the right result is?

To say that something is the right result is to say we acknowledge it.—There are all sorts of ways of following the rules. The experiment does not show the right result. And to show that something is the right result is not showing that it is the result I get and also something over and above it.

Suppose that we make enormous multiplications—numerals with a thousand digits. Suppose that after a certain point, the results people get deviate from each other. There is no way of preventing this deviation; even when we check their results, the results still deviate. What would be the right result? Would anyone have found it? Would there be a right result?—I should say, "This has ceased to be a calculation."

We are used to the symbol "$\triangleq$", not in pure mathematics (at least, not in this way), but in the application of mathematics to physics. But one could introduce it into pure mathematics. One could say that such-and-such a number multiplied by such-and-such another number is roughly so-and-so, where "roughly" is a mathematical symbol. In this new kind of mathematics we might say that a child is correct if his answer is 3 more than mine, but if he got 5 more, it would be incorrect.

As regards Watson's suggestion, I might say that the multiplication of 136×51 makes me adopt a new rule. I proceed from certain rules, and I get a new rule: that $136 \times 51 = 6936$.

Suppose that you count a number of objects. How do you know that you have not left out a numeral in your counting?

Turing: You don't know.

Wittgenstein: Well, you may or may not get into difficulties. But we don't get into difficulties. The fact is, all grown-ups count alike and do not, when asked to count objects, constantly hesitate and say, "Now did I leave out a number in counting?"

Suppose that from now on, when we were told to multiply, we all of us constantly got different results. Then I suppose we should no longer call this calculation at all. The whole technique (for instance, of calculating floor boards) would lose the character of a calculation. We would then no longer in fact have a right or a wrong result.

The whole thing is based on the fact that we *don't* all get different results. That's why it was so absurd to say $12 \times 12 = 144$ may be the wrong result. Because the agreement in getting this result is the justification for this technique. It is one of the agreements upon which our mathematical calculations are based.

Wisdom: One might ask whether one knows by calculation whether or not one has got the right result. For do I know that $2 + 2 = 4$ by intuition or is it a question of taste? When the results are regular there is less inclination to call it a question of taste.

Wittgenstein: It is not a question of whether it is a question of taste, but of what is regarded as a question of taste. If the Chinese multiply differently from us, one can say that it is a question of taste whether one multiplies in our way or in theirs. But it is not a question of taste whether Lewy says that $2 + 2 = 4$ or that $2 + 2 = 3$.

XI

Turing thinks that he and I are using the word "experiment" in two different ways. But I want to show that this is wrong. That is to say, I think that if I could make myself quite clear, then Turing would give up saying that in mathematics we make experiments. If I could arrange in their proper order certain well-known facts, then it would become clear that Turing and I are not using the word "experiment" differently.

You might say, "How is it possible that there should be a misunderstanding so very hard to remove?"

It can be explained partly by a difference of education.

Partly by a quotation from Hilbert: "No one is going to turn us out of the paradise which Cantor has created." [1]

I would say, "I wouldn't dream of trying to drive anyone out of this paradise." I would try to do something quite different: I would try to show you that it is not a paradise—so that you'll leave of your own accord. I would say, "You're welcome to this; just look about you."

One of the greatest difficulties I find in explaining what I mean is this: You are inclined to put our difference in one way, as a difference of *opinion.* But I am not trying to persuade you to change your opinion. I am only trying to recommend a certain sort of investigation. If there is an opinion involved, my only opinion is that this sort of investigation is immensely important, and very much *against the grain* of some of you. If in these lectures I express any other opinion, I am making a fool of myself.

Take our example of the enormous multiplication with numbers of 1000 digits.[2] Here it seemed, queerly, as if calculation and experiment were getting closer and closer.

The answer was: Quite possibly the best you can get is that Mr. So-and-so has arrived at this result. And it looks as if this means we cannot reach the mathematical result, but this is the nearest we can get.

I should say that if it was a mathematical proof, God didn't know more than any one of us what the result of the calculation was.

"For us human beings, the best thing we can arrive at, the nearest we can get, is that we always get it, or someone who had a lot of experience always got it." As if only God really knew. —Turing suggested this, and that is just where he and I differ. Actually there is nothing to stop us postulating that your result

1. From David Hilbert, "Über das Unendliche". "On the Infinite", a translation of this essay, appears in *Philosophy of Mathematics: Selected Readings,* ed. P. Benacerraf and H. Putnam (Englewood Cliffs, 1964). The quoted sentence occurs on p. 141. Wittgenstein may have quoted from Hardy, "Mathematical Proof", p. 5.

2. Lecture X.

is right—so that in future all your children will have to copy what is written on that blackboard. And then it is right.—There is nothing there for a higher intelligence to know—except what future generations will do. We know as much as God does in mathematics.

What I say doesn't contradict the statement that Mr. So-and-so's calculation may play the role of an experiment. And this in no way lies in what he does with the chalk when he multiplies—in anything that happens at the time.

There may be various misunderstandings: (1) Turing said that the characteristic of an experiment is that we are interested in the result. But of course that is not true. Chopping wood is not an experiment just because one is interested in the result; or washing one's hands. It *can* be an experiment. So can doing things with chalk on a blackboard—it can be thousands of experiments. (2) If a thing is regarded as a calculation, then it is thereby *not* regarded as an experiment—the two things are contradictory.

I once said: A calculation could always be laid down in the archive of measurements. It can be regarded as a picture of an experiment. We deposit the picture in the archives, and say, "This is now regarded as a standard of comparison by means of which we describe future experiments." It is now the paradigm with which we compare.—It is as if somebody said to me, "How do you write a capital F?" I write one. Then he declares, "From now on this is *the* capital F" or "All capital F's shall be described in terms of this one, as more or less deviations from it."

"If multiplying is an experiment, it is rather queer that we should ever make this experiment."—It would be made in a psychological laboratory, if anywhere.

Making this picture of so-and-so's experiment and depositing it in the archives—you might call doing it an *honour.* We should only do it if the experiment was of a very peculiar kind. For instance, it must be connected in certain ways with what is likely to be the result of other similar experiments. That I should take this procedure as the standard procedure means a whole lot: that it is the right procedure and at the same time removed from possible tests—this is bound up with a lot of opinions of mine about what's going to happen.

Turing: The difficulty is that there is not a finite number of multiplications. You can only put a finite number of multiplications in your archives; and when I do a multiplication which is not in your archives, what then?

Wittgenstein: Well, what then?—This is like counting to a number which has not been counted to.

Now what is it that we are going to deposit in our archives? We might say, "We are not going to deposit single multiplications, but only general rules."

But let us go into this question. We have the metre rod in the archives. Do we also have an account of how the metre rod is to be compared with other rods? There might be a point sometimes in putting an account—say, a picture—of the way in which we compare them; or instruments used for this purpose.

Couldn't there be in the archives rules for using these rules one used? Couldn't this go on forever?

But this has nothing to do with the fact that the number of multiplications is infinite. In fact, that it has no connexion with it is an important point. The idea that it is connected with it comes from the idea that the examples, being infinite, are too numerous to go into the archives.

We might put into the archives the multiplication table. It will be put in to keep this technique. Anyone who wants to know how people do it can go in and find out: "Yes, that's how people do it."

Or we might put into the archives just one multiplication—as a paradigm for the technique. As we might keep a paradigm of pure colour.

Why keep

```
   465
   159
  ----
  ....
 ....
 ...
 -----
 .....     ?
```

It would make sense to do this if everyone knew from it how to multiply in other cases. (Compare induction.)

Or we might keep in the archives a general description of multiplying.

But to go back to Turing's difficulty: "an infinity of multiplications". We *might* say every new multiplication made is a new rule made.

Then why make multiplications at all?

Supposing we do a multiplication: the use of this is that we aren't willing to recognize a rule of multiplication *unless* it can be got in a particular way. For instance, we do not accept the rule that $1500 \times 169 = 18$; we should not call that a multiplication.—The way in which it can be got we accept or acknowledge as a *proof* of it.

Turing: If we were only concerned with multiplications up to 10, we could put them all in the archives; but as it is, the case is quite different.

Wittgenstein: Yes, and it is important to see that the two cases are different.

Isn't there an infinite possibility of examples of putting metres end to end?—I might or might not wish to give examples. But it may be entirely unnecessary to give examples.

The point is this. Suppose I put into the archives a general rule and a few examples; and you now give a new example. This might be a new rule—and we need not put this into the archives, but we might do so. The fact is that we *recognize* it.—To say that it is infinite doesn't mean that there is such a large number that we can't get it into the archives. The fact that there are or are not an infinite number of examples is entirely irrelevant.

In the case of the vast multiplications, it looked as if we had something that was a hybrid between an experiment and a calculation—or that *faute de mieux* we had to put up with an experiment.

Suppose I ask Wisdom to multiply two very large numbers, and later ask him what the result was. He says, "I had such an awful headache, I don't know really, but I got so-and-so." You might say, "There you are. We have now got the result of an experiment made under the wrong conditions."

But if he says, "This was what I got"—this is not the mathematical proposition. How do we pass from this to the mathematical proposition: "So-and-so times so-and-so *is* so-and-so"?

It has been said: "It's a question of general consensus." There is something true in this. Only—what is it we agree to? Do we agree to the mathematical proposition, or do we agree in *getting* this result? These are entirely different.

What is it they must agree in? They agree in *getting* this. They may agree in saying "I got so-and-so"—in finishing up with the same number, etc. But not that this is the answer. There isn't as yet such a thing as 'the answer'—because there isn't yet a technique. So far, ". . . times . . . *is* . . ." doesn't mean anything: there isn't yet a mathematical proposition. They agree in what they do.

Mathematical truth isn't established by their all agreeing that it's true—as if they were witnesses of it. *Because* they all agree in what they do, we lay it down as a rule, and put it in the archives. Not until we do that have we got to mathematics. One of the main reasons for adopting this as a standard, is that it's the natural way to do it, the natural way to go—for all these people.

Wisdom: The idea that a mathematical calculation is an experiment is connected with the idea that a mathematical proposition tells us how people use language. For example "$2 \times 2 = 4$" means "When people are asked what 2×2 is, they generally answer '4'." If one thinks that mathematical propositions are concerned with how people use language then one is inclined to think that one first experiments with oneself and then from that experiment predicts how other people will use language. Similarly, one may feel inclined to say that I find out whether this picture is good or whether this book is red by experiment; for I first look at it myself and then from that predict what other people will say about it.

Wittgenstein: I will talk about this some time. But honestly it isn't on the main road at the moment. But I am not shirking it. However, I will say this: it is true that when we have been conditioned in a certain way we react in certain ways, and that we may conduct experiments to find this out.—Suppose that a schoolteacher makes statistics of the progress of his class in mathematics. He might write down in his diary under a certain date: "75%: $6 \times 6 = 38$", meaning that 75 per cent of his class said that

6×6 was equal to 38. And a year later he might write "99%: $6 \times 6 = 36$". In that case the mathematical sign "=" would be being used in the way Wisdom has suggested. But in fact mathematical symbols hardly ever are used in that way.

But we must get back to the main question. We have all of us worked out certain multiplications. And actually there are no disagreements about the result of a multiplication—so that we don't know what to believe because we always have a headache, or all the people get different results. This hasn't happened; that is immensely important. But we can imagine that when we do a new multiplication it does happen. What about right or wrong in this case? Half the people do it one way, half another way. Our last resort is "But don't you see?", and this doesn't change it.

So what are we to say? That none of us knows what the true answer is, or that we have not decided what the true answer is?

What is the sort of difficulty that has arisen? Is it that some of them believe it is one thing and some another?—We can't say that some follow the rule in the right way, and others in the wrong way—for we don't know what this means.

Turing: One might say that the two lots of people had seen two different analogies between this case and the multiplications in the archives, and therefore get different results.

Wittgenstein: Ah, I knew you would say that. But what is wrong with that answer can be seen by asking: *which* two different analogies? You can legitimately say, "They see two different analogies" in certain circumstances: if you can go on to say, "*namely,* these people see this analogy, the others see that analogy."—You can't have an internal relation unless you have both terms already.

You give an explanation of analogy: "The one sees it as so-and-so, the other as so-and-so." You must have an expression to describe the analogies different from just the result of that new calculation. For your expression of the analogy might be used differently. They might see the same analogy and get different results, or different analogies and the same result, or no analogy.

You might say, "I know exactly why these two people differed. They saw different analogies"—and clear up why they got different results. Compare: "The one saw the man *as a cross,* the other *as a pentagon.*" It is then a hypothesis that they did see different analogies—this could be found out by experiment.—But it is not clear at all how the symbols of cross and pentagon are to be used.

Suppose that I write a series, and ask two people to continue it, and they write down different results. "The reason is that they see different analogies." "Two analogies" would be like what we could express by two different algebraic formulae. The explanation makes sense only if there is a test for their seeing *these* two different analogies.—Otherwise: the one writes this, the other writes that. Bringing in two analogies won't help you; it is only bringing in two new symbols which can be applied in different ways.

So what about our case: this new calculation and these people disagree. What are we to say?—Shall we say, "Why aren't our minds stronger?" or "Where is an oracle?" But is there anything for it to know? Aren't you right—or wrong—as you please?

Turing: We'd better make up our minds what we want to do.

Wittgenstein: Then it isn't a message from God or an intuition, which you pray for, but it is a *decision* you want. But doesn't that contradict the idea of an experiment? Where is the experiment now?

Turing: I should probably only speak of an experiment where there is agreement.

Wittgenstein: Don't you mean that in that case the experiment will show what the rule is?

The fact is that we all multiply in the same way—that actually there are no difficulties about multiplication. If I ask Wisdom to write out a multiplication and get the result, and he tells me, then I am perfectly certain that that's the right thing, the adopted thing.

I can find out by this experiment what the rule is. But does this make the rule the result of an experiment?

Watson: The reason why one thinks that in all such cases of agreement and disagreement there must be a right and a wrong is that in the past there have been mistakes in mathematical tables, with the result that if one used these tables when building a bridge, it would probably fall down.

Wittgenstein: The point is that these tables do not by themselves determine that one builds the bridge in this way; only the tables together with a certain scientific theory determine that.[3]

We might have mistakes—as we now say. Under what circumstances should we say that for five hundred years people have had the wrong log tables? that is, with the result that we would now change them.—Although one cannot say that there are no such circumstances, yet the case is somewhat similar to the suggestion that "$12 \times 12 = 144$" is wrong.

"They all reasoned wrongly."

This may mean all sorts of things.—If you don't know the special case, you don't know at all what they did—you don't know what 'reasoning wrongly' means.

We might mean, for example, "If only we had said to them so-and-so they'd have seen it."—But often this isn't so. Think of disputes about transubstantiation. It is not true that if someone had said to Luther and Zwingli that the meaning of the word 'wine' is the method of its verification, they would have said, "Oh, now I see" and stopped arguing. On the contrary they might have killed you—and perhaps rightly. That is, I am not saying that they would be behaving stupidly.

Suppose we say just: First of all, they reason differently from us.

Now why "wrongly"?

"They didn't see these connexions"—Does this mean they didn't talk about them?

This question of "they reasoned wrongly" is immensely complicated and must be treated another time. But it is connected with Watson's point about mathematical tables. For when we find such mistakes, we do not say, "Well, now we do it differently"; we say that they made mistakes.

3. This sentence is only in B, where it was included in Watson's remarks. It seems more likely that it was part of Wittgenstein's reply.

Next time I hope to get on to counting. I want to compare "We count the people in the room in order to find out their number" with "We count the number of permutations of such-and-such things." For the latter case looks as if it were a case of mathematics being applied to mathematics itself.

XII

These discussions have had one point: to show the essential difference between the uses of mathematical propositions and the uses of non-mathematical propositions which seem to be exactly analogous to them.

Mathematical propositions are first of all English sentences; not only English sentences, but each mathematical proposition has a resemblance to certain non-mathematical propositions. —Mathematicians, when they begin to philosophize, always make the mistake of overlooking the difference in function between mathematical propositions and non-mathematical propositions.

Hence we want to see the absurdities both of what the finitists say and of what their opponents say—just as we want in philosophy to see the absurdities both of what the behaviourists say and of what their opponents say.

Finitism and behaviourism are as alike as two eggs. The same absurdities, and the same kind of answers. Both sides of such disputes are based on a particular kind of misunderstanding—which arises from gazing at a form of words and forgetting to ask yourself what's done with it, or from gazing into your own soul to see if two expressions have the same meaning, and such things.

In a most crude way—the crudest way possible—if I wanted to give the roughest hint to someone of the difference between an experiential proposition and a mathematical proposition which looks exactly like it, I'd say that we can always affix to the mathematical proposition a formula like "*by definition*".

"The number of so-and-so's is equal to the number of so-

and-so's": experiential or mathematical. One can affix to the mathematical proposition "by definition". This effects a categorial change. If you forget this, you get an entirely wrong impression of the whole procedure.

The "by definition" always refers to a picture lying in the archives there.—If we forget this, we get into one queer trouble: one asks such a thing as what mathematics is about—and someone replies that it is about numbers. Then someone comes along and says that it is not about numbers but about numerals; for numbers seem very mysterious things. And then it seems that mathematical propositions are about scratches on the blackboard. That must seem ridiculous even to those who hold it, but they hold it because there seems to be no way out.—I am trying to show in a very general way how the misunderstanding of supposing a mathematical proposition to be like an experiential proposition leads to the misunderstanding of supposing that a mathematical proposition is about scratches on the blackboard.

Take "20 + 15 = 35". We say this is about numbers. Now is it about the symbols, the scratches? That is absurd. It couldn't be called a statement or proposition about them; if we have to say that it is a so-and-so about them, we could say that it is a rule or convention about them.—One might say, "Could it not be a statement about how people use symbols?" I should reply that that is not in fact how it is used—any more than as a declaration of love.

One might say that it is a statement about numbers. Is it wrong to say that? Not at all; that is what we call a statement about numbers. But this gives the impression that it's not about some coarse thing like scratches, but about something very thin and gaseous.—Well, what is a number, then? I can show you what a numeral is. But when I say it is a statement about numbers it seems as though we were introducing some new entity somewhere.

I gave an example of how a calculus can be introduced—to help us plank the floor. The children are taught to make certain arithmetical statements. Are we to say these are statements about numbers?—Here you immediately see something queer. I said it was tempting to imagine a mathematical statement having a function similar to an experiential statement of the same structure. We've called such a statement as "20 + 30 = 50" a statement about numbers; this seems like saying: it's a statement about apples if I say, "Take 5 groups of 5 apples and you get 25 apples." [1]

Now is "20 apples plus 30 apples is 50 apples" about apples? It might be—saying apples didn't join up. But of course it may be a mathematical statement.

Might we not put all our arithmetical statements in this form—statements in which the word "apple" appears? And if you were asked what an apple was, you would show the ordinary thing we call an apple.

Take the case of *Principia Mathematica,* where we use p and q and r. Couldn't one take instead of p, say "it rains", and so on? If you liked you could take "it rains" to be one letter, a very complicated letter.

This is important when we come to consider generality. We can have a perfectly general proof, using "it rains" instead of p. I say "We *can* have": the point will be whether we *only* prove that either it rains or it does not rain, or whether we also have proved at the same time that either I am going to London or I am not going to London.

Similarly when we prove that 20 apples + 30 apples = 50 apples, we may have thereby proved also that 20 chairs + 30 chairs = 50 chairs or we may not.—What is the difference between proving it for apples alone and proving it for chairs, tables, etc? Does it lie in what I write down? Obviously not—nor in what I think as I write it. But in the use I make of it.

"20 apples + 30 apples = 50 apples" may not be a proposition about apples. Whether it is depends on its use. It *may* be a proposition of arithmetic—and in this case we could call it a proposition about numbers.

1. (From "I gave".) The four versions of this passage are quite different; material from all of them has been used.

You might ask, "Isn't there something queer about this? How could all this have changed what it is about?" But that is how we use the phrase "a statement about numbers". As soon as it's applied in a certain way, we say it's about numbers.

And a *discovery*—"627 + 324 = . . ."—could in one case be called a discovery about apples, in the other case a discovery about numbers—according to what we do with it.

The change from 'being about apples' to 'being about numbers' is an entirely different kind of change from:

"Lions are four-legged"—this is about lions.

"Elephants are four-legged"—this is about elephants. Or from 'being about apples' to 'being about pears'. In fact it is the same proposition when it is about numbers and when it is about apples, only it is used in an entirely different way. When it is put in the archives at Paris, it is about numbers.

You'd expect, if it is about numbers, that you've made a discovery in a new realm. But it is not in a new realm at all. You have made something entirely different.

I said that the whole trend of these discussions was to show the difference between mathematical propositions and experiential propositions which look exactly like them. Now consider the case of counting in ordinary life, for example, counting the number of people in this room, as compared with counting in mathematics, for example, counting the roots of an equation.

If we are asked why we count, we are tempted to say that we count in order to find out the number of things.

This is like: "We weigh in order to find out the weight of things." But there may be something fishy about this—and may not. "Knowing the weight" sometimes comes to "knowing what happens when you weigh". But we are liable to confuse the case where what we call the weight is by definition what we find out with the case where it is not. For instance, you may want to find out what the weight will be as felt, or whether he will be able to lift it. I might *define* the weight by one method of weighing, and then say that you can also find the weight by some other method. But this is quite different from the case where one defines the weight as the result of weighing it on the scales and then finds out the weight by putting it on the scales. (The analogy here with: "by definition".) [2]

"Counting because we want to know the number of things." Doesn't the same apply here?

Is it always the case that we weigh in order to find out the weight of the object? No; you may weigh, for instance, to find out whether your balance is correct. The same applies to measuring—"in order to find the length". We might measure to find out if the foot rule is all right. We do exactly the same things, but this need not be an experiment to find the length. Similarly in many cases you may count to find the number, but it may not be done for an experiment at all. One may count in order to play a game.

What is counting? Pointing to things and saying "1, 2, 3, 4"? But I need not say the numbers: I could point and say "Mary had a little lamb" or I might whistle "God Save the King" or anything. —But normally the process of counting is used in a different way, whereas "Mary had . . ." is not used in this way at all. If you came from Mars you wouldn't know.

Now what sort of application of one's actions and words would you call finding out the number, and what sort would you not call finding out the number?

Turing: If, for instance, you wanted to give us each four buns and you pointed to each of us in turn and said "One, two, three" and then went and bought twelve buns, then we should say that you had counted.

Wittgenstein: Yes. Now what I am driving at is the difference between counting the people in this room and counting the points of intersection in the pentagram.

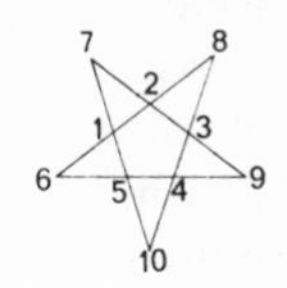

"The pentagram" is the name of this figure.

2. Material from all four versions has been freely combined in this paragraph.

Is there a difference between the use of the sentence "There are ten people in this room" and "There are ten points of intersection in the pentagram"?—The latter is a mathematical proposition and the former is not. Why? "Well, the one talks of people, the other of points and lines." But that is not an essential difference. We have just said that mathematical propositions might quite well be expressed in terms of people, houses, or what not. The word "men" may come in and it may still be mathematics; and the word "lines" may come in and it may *not* be mathematics.

For instance, "There are 13 lines here" ||||||||||||| seems not to be a mathematical statement. Why not? What is the difference between counting these lines and counting the points of intersection here  ?

Can we say, "By definition there are ten points of intersection in the pentagram?"

We might be inclined to call counting always an experiment; it is very similar to measuring. Suppose I had a 'counting rod': this is a rod with numerals marked on it—not necessarily at the same distances. The rod is used for counting: for instance, when I want to count men, I place a man at each numeral, and I read off the numeral beside the last man.

Now we're inclined to say *measuring* is an experiment; and counting seems just as much so.

So it seems that in the case of the thirteen lines we have made an experiment and got an experiential, non-mathematical result —and that in the case of the pentagram we have made an experiment and got a *mathematical* result. Why is this?

Gasking: In the latter case, one counts not the number of intersections of this particular pentagram but the number of intersections of *the* pentagram.

Wittgenstein: Yes—but how misleading that is! For what does counting the number of intersections of *the* pentagram consist in, as opposed to counting the number of intersections of this figure? Isn't it queer that you distinguish objects and that you say that one is counting different things in the two different cases? Isn't this most misleading? How do you know what I have counted? I have just written down these numbers; how can you tell what I'm counting? By my merely writing down the numbers you can't tell.

While I was actually counting, I may have been thinking of the chalk marks and not about *the* pentagram—but that would not mean that I had not counted the intersections of the pentagram. The difference is not in what goes on at the time of counting.

If I say merely, "The figure I have drawn here has ten intersection points", you would know at once that it was not a mathematical proposition. If I say, "The pentagram has ten intersection points", that at once suggests the mathematical use.—But of course the sentence "The figure I have drawn here . . ." may be used either mathematically or non-mathematically.

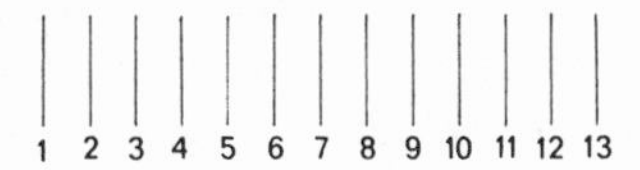

(1) "This figure has thirteen lines."

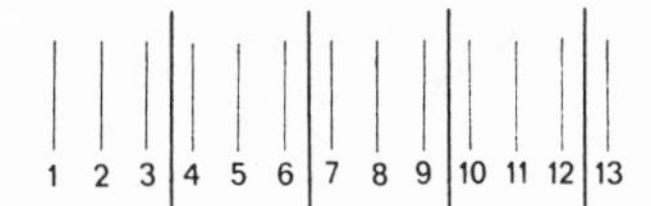

(2) "This figure has thirteen lines."

(1) we are not inclined to call a mathematical proposition, (2) we are. Why is this?—One would be inclined to say that here it means that $(4 \times 3) + 1 = 13$. If I had some sort of name for this, it would be in the same case with the pentagram.

Can't we have measuring which is not an experiment?—where it is like counting the intersections of the pentagram.

Take the case of measuring the standard metre rod with a foot rule. Isn't this an experiment? If I saw someone doing this, I wouldn't know what he was measuring.—He could find out how to express metres in terms of feet; or how long the foot was.

But suppose we place a yardstick against a metre rod, glue the two together and then cut off the metre rod where the yardstick ends so as to make one stick of them. Can we say which is being measured by which?

Turing: Probably the yardstick is being measured by the metre rod, as the metre rod is cut off where the yardstick ends.

Wittgenstein: Oh, come now—then we will cut it somewhere in the middle. Now you cannot tell which is measured by which.

The point is this. What is this experiment which the man has done? He has produced this double stick, but what then?

Turing: He might use it for digging in the garden or anything.

Wittgenstein: Yes, certainly he might. That is one application of it. But there is a very obvious application of it.—The result is a rod with metres on one side and feet on the other—and he can now use it to transform one into the other. This is a table or a rule.

Here we have an analogy with writing the numerals on a pentagram. No one would say that when a man makes a stick like this he is making a measurement. For although he does the same as someone who is measuring, namely put two things side by side, the *use* of it is quite different.

Similarly, you could use the pentagram for counting; and you could not even say which I've counted, the numerals or the points of intersection.

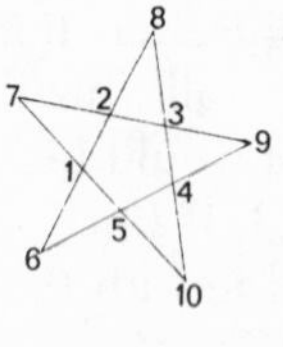

Suppose I say, "Here I've counted; I've said 'One, two, three' and so on." But what have I got at the end? A mathematical proposition.—What makes it into a mathematical proposition?

You might call it two ways of counting glued together. We could have had one way of counting by putting people on the crossing points of the pentagram and another way of counting by assigning numerals up to ten persons. What looks like counting, in the case of the pentagram, is a way of correlating these two ways of counting. [A rule is made.]

The point is this. If you ask, "Did I make an experiment when I counted?"—no more than when I glued the metre rod and the yardstick together. I did something like measuring, but I didn't measure anything. In both cases, I produced something new—a new paradigm.

Isn't this what always happens in mathematics when we count entities, say, the roots of an equation?—The crucial point is: he *glued* them together, he made *one* instrument out of them.—You can't say you've necessarily counted the roots by means of the numerals, or vice versa. You could perfectly well count by means of the roots of equations, and then introduce cardinal numbers afterwards.

Suppose that someone writes down the series of cardinal numbers and always underlines the primes. He notices that in many cases the primes come in pairs.—Or suppose he did this:

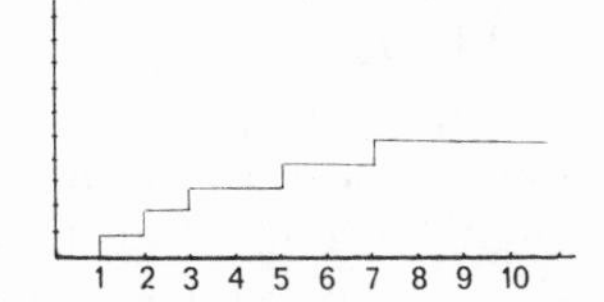

He draws a line, going up one whenever he comes to a prime.—Say now I go on for 500 places, and try to put a continuous curve through, which as near as possible connects these points. Suppose I found a bit of a parabola going pretty nearly through these points when I have them up to 1000.—This resembles very closely what is done in physics.

What is the difference between the curve I've drawn and the one the physicist draws? Have I in a sense made a mathematical experiment here, drawing the diagram?

Suppose I get a parabola and write down the equation, and say, "That goes through all or nearly all the points up to 10,000." —Or suppose someone had found a continuous curve which he *proved* went through the whole lot—increasingly satisfactorily. Is this a proof that the mathematical experiment had to have such-and-such a result? Does this make it a mathematical statement?

Turing: That depends upon how it is phrased. For instance, if one says that the logarithmic curve 'fits quite well' then that is not a mathematical statement.

Wittgenstein: Now why is it not a mathematical statement? What would make it into a mathematical statement?—Well, the phrase "quite well" is not a mathematical symbol. "Approximately" sounds better, though it means just the same. No technique has been arranged for the word "approximately".

Turing: When one says that a thing fits quite well one means that it gives one a certain satisfaction. But if one said that it was never more than 2 out that would certainly be a mathematical statement.

Wittgenstein: But suppose a man tells me that such-and-such a formula fits quite well up to 1000; and when I ask if he would say that it fitted quite well if it deviated by so much, he says "No"—then immediately we have a mathematical statement.

Turing: I should say that the only information which he is then conveying to me is that if the curve had deviated by so much, it would not have satisfied him.

Wittgenstein: "This fits quite well" may be used strictly as a statement that I am satisfied with it. For instance, a man might want a curve drawn roughly through these points for a wallpaper pattern; and when I draw a curve for him, he might say, "Yes, that fits quite well."—But let's put that case on one side.

Suppose now that we define 'fits well' exactly, so that it means, say, "not deviating by more than 3 from any of the points". If I say, "The formula so-and-so makes a curve fitting well up to 100"—does this mean it *must,* or it does? The 'must' seems to be in a sense the criterion for the mathematical quality.—One can put the case in a still more misleading way by saying, "Well, so far you've only shown they fit well up to 1000, not that they must. But if the proof is given, you've shown they *must* fit well."

Suppose we take:

$$1 : 7 = 0.142857\underbrace{142}\ldots$$

Have I shown that these figures ↗ *must* come here when I've done the division simply? or have I shown they must come there only when I've *proved* the recurrence? (Imagine that the remainders had been written down, but it hadn't been pointed out that we'd had a remainder the same as the dividend—or that the 1 with which we started recurred.)

Lewy: By dividing, you've shown that those figures must come there.

Wittgenstein: Yes. You have done the whole division. If the calculation is correct you've shown 142 *must* come there. The general proof doesn't make it clearer.

The same applies with a curve. If you have shown that up to 1000 the curve fits well, then you have shown it *must* fit well. A general proof doesn't add anything to the logical necessity of that.

Suppose I say, "I've found that the prime numbers often come in pairs." Is this the result of an experiment?—Here it looks just like an experiment. I didn't know what the result would be, and I found out by going through some divisions.

Wisdom: In this case you have shown it not by experiment but by proof.

Wittgenstein: Yes—but why do we say this here?—There is no difference between showing that they come in pairs and showing that they *must* come in pairs, just as there is no difference between showing that 17 is a prime and showing that it *must* be a prime.

It looks like getting readings and then connecting them. But which is the primary thing here? (As with the measuring rod, or counting the intersections of the pentagram.) It looks as if one necessarily comes before the other, as if one necessarily measures the other. And in an experiment that is so. But here I have simply coordinated a certain formula and a certain technique: the primes and the cardinal numbers.

I will just suggest the following problem, and then talk about it next time. It has often been said—and there is something true in it and something absurd—that a mathematician sometimes makes what one might call experiments, and then proves what he has found out by experiment. But is this true? Is not the figure itself—the curve or the division—a proof?

A child might divide 1 by 7 up to 100 places and never notice that the remainder recurs. He might then ask, "Is it accidental that these six have recurred in the answer, or *must* it happen?" But is this the right way to ask the question?

Turing: One might ask, "Can I come to understand it better?"

Wittgenstein: Yes—but that is the whole point. What does one in such a case call "coming to understand it better"?

But when he says, "Must it happen?" he means, "Can I give up dividing and just go on repeating the answer instead? Or if I do that may I go wrong after a certain point?" It doesn't mean that what the mathematician calls the result of the experiment will later be proved. This *HAS* been proved. Something else hasn't been proved, but this hasn't been shown by experiment either. He made calculations and then said, "Perhaps this always must happen." As far as the experiment goes, he might say, "Perhaps I'll go on writing these numbers over and over again."

What he is inclined to say he has found out by experiment he has found out by proof. The way he goes on to make a kind of conjecture, something like "Perhaps it can be proved that this must always go on", sounds as if he had made an experiment.

XIII

I tried to make the distinction between ordinary counting and counting in mathematics, for example, counting the intersection points of the pentagram.—If we treated counting the intersection points in the pentagram as an experiment, what is the result?

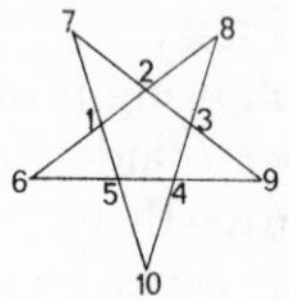

You might ask, "What is it we've counted? Is it the intersection points of 'the pentagram' or the intersection points of this figure I've drawn here?" What I've done couldn't be called one or the other by watching me write these things on the blackboard. So it couldn't be told offhand whether I am expressing a mathematical or an experiential proposition.

We said last time that in mathematics it seemed as if we made certain experiments, the results of which seemed to fit a hypothesis, which later on we proved.

Take the case of the man who divides 1: 7 = 0.142857142 . . . When he notices the recurrences, he says, "Well, I suppose this will go on for ever." Then he finds a proof of this in the recurrence of the dividend in the remainder.

If you say he first made a certain hypothesis or assumption and later proved it—this seems in a way to contradict all that we have said before. Suppose we say, "He had for a time *believed* that they would always recur and then found that that they *had* to recur." This puts the proposition he *believed* on a level with experiential propositions. Thus Professor Hardy in an article on "Mathematical Proof" in *Mind* says that one can believe a mathematical proposition, as opposed to knowing it.[1]

The point is: what is the relation between that unfounded belief and the proved proposition that it *must* recur?

Suppose you had learned to multiply up to a certain point. Ordinarily in learning the technique of multiplication one is given certain rules and examples—and there comes a point where you're *encouraged* to go on. This is an infinite technique or an $\aleph_0$ technique. But suppose you'd been taught to multiply in such a way that at a particular point the teaching stopped. Things simply came to a standstill. How could this be? Well, there might be many things. For instance, you never heard of anyone multiplying further. And perhaps one day the teacher says, "Well, today we will do the last multiplication: $10 \times 10 = 100$."—In this case what would it be like for someone to believe that $10 \times 11 = 110$? Suppose someone said that; what would it mean to say that his belief might be proved right or wrong? Wouldn't this be queer?

1. Vol. 38 (1929), p. 4: "When we know a mathematical theorem, there is something, some object, which we know; when we believe one, there is something which we believe; and this is so equally whether what we believe is true or false." Cf. also p. 24: ". . . my invincible feeling that, if Littlewood and I both believe Goldbach's theorem, then there is something, and that the same something, in which we both believe, and that that same something will remain the same something when each of us is dead and when succeeding generations of more skilful mathematicians have proved our belief to be right or wrong."

Well, we have come to a certain point and you can go on as you like. Similarly, I may ask, "What do you believe is the continuation of this: ⌒ ?" To say this offhand would mean nothing. You could continue in any way you please. If he said, "The continuation of it is ∽ ", isn't this like making a decision?

Similarly with "believe that they will recur". Offhand we might say, "It means nothing at all just to say we believe they will recur."

You might believe two totally different things. The phrase is misleading: "will recur" as normally used is a mathematical phrase. It is not a temporal expression; it doesn't mean "will recur with most people" or "will recur in half an hour" or anything like that. It means something totally different.

You could mean: (1) "If people do this, it will always recur in their writing", or "People being conditioned in such-and-such a way . . ." *Or* (2) "I believe that if they calculate correctly it will recur", or "It will be right if it recurs."—But this is a queer phrase. One might say, "I believe it won't recur and that after dividing to 1000 places they will be so tired that they will miss out figures"—but one says, "I believe it will recur if the calculations are correct." This is very queer, and totally unlike believing that I shall be hungry tomorrow. Whether it is calculating correctly or not depends in a way on your decision. You might say, "I believe we'll have to adopt *this* as the correct rule."

Watson: It is more complicated than that, because one already has a rule which seems to determine what will recur.

Wittgenstein: Yes, we have a rule—and the point is that now we have to make another rule, of which we say it *agrees* with the first rule.

We have then a rule for dividing, expressed in algebraic or general terms,—and we have also *examples.* One feels inclined to say, "But surely the rule points into infinity—flies ahead of you —determines long before you get there what you ought to do."

"Determines"—in that it leads you to do so-and-so. But this is a mythical idea of a rule—flying through the whole arithmetical series.

We might say we have here a decision: "I can now use a different technique. I think it'll do"; and then . . .

"I think it'll do"—that is, try it.

We are in the midst of a large number of queer puzzles.

Suppose we divide to thirty places, and someone says we needn't divide but only copy after the first six places. You could say, "How on earth can a logical proof be shortened? Either you have to divide or you needn't divide. If we just copy out the figures, then we do not divide in the old way any more."—If we say the proof that the first thirty places are so-and-so is given by dividing, then isn't it queer that this can be shortened? Either this is *the* proof—but then it must be given: how can we leave out a part of the proof and still get the proof? Did we at first do something superfluous? If it's a *logical* proof, then that *alone* should justify the conclusion. To say something else can justify it seems to make *logic* rather arbitrary.

"Surely you can see we needn't go on"—all I know is that you won't go on.

If you leave off dividing and do something else instead, what is it that you have shown?

Now Watson said we believe that copying this will do. Will do for what? What trick is it supposed to do?—Someone might say "Judge the result of the division by its practical application. If we distribute nuts to people, then this new process will work well." But this is not what Watson meant. What is to be the criterion for its working well? So far only: that it gives the same result as a normal division.

To see the connexion between copying and dividing, let's suppose we adopted a *different* rule. For example: Up to ten cycles the cycles were all of this kind. The next ten cycles are different—they each have a 1 in front of them; the next ten have two 1's, etc.—You may say, "This obviously won't do." There would now be what we would call a disagreement. There would be complete agreement in the first ten cycles; then there would be disagreement in our minds—up to the point where we leave off dividing.

We could reconcile this disagreement in many ways.—Compare: Suppose I said a hundred times "One, two, three, four, five, six, seven". What would be the criterion that I had left out something? Well, we adopt many criteria—people hearing me, etc. Yet we could say that without knowing it we leave out something whenever we do that.—Similarly, we could perfectly well adopt a means of explaining the disagreement: we get more and more tired, and leave out more and more digits. But to say this would be very inconvenient.

The fact that my rule for putting in 1's seems so utterly absurd, already shows that it would mess up the whole use of mathematics, just as it would if we constantly assumed we had left something out.

Similarly, if I ask you where you've been, there will be long stretches when you can't remember at all, and no one else can either. We might then say, "I suppose you vanished." Or we could assume physical bodies always disappeared when no one looked at them. To describe things in this way would be quite possible. But it would make things extremely awkward. Just as it would be awkward if we constantly assumed there were animals in this room in places where no one could possibly see them.

We could even teach people to divide numbers up to the point where the remainder recurs, and that the division then goes on by copying—without ever teaching them to divide further. Division would be defined in this way. There would then be no theorem that it recurs. *This* would be division now. The child doesn't ever learn how to go on: there's no such thing.

Let us first have division taught in this way. And then I ask you: why should we even copy it out? Why not copy it out with a variation each time? That would simply be a new calculation.

Watson: In a sense it is a practical question. For instance, when making a table of squares, one first multiplies the numbers and then checks—for example, by seeing whether the last figure can be the last figure of a square. One uses this process for making up the tables, and the tables are used for practical purposes.[2]

Wittgenstein: If you say, "This will do" you might mean: this

2. It is not certain where Watson's remark belongs. B has it after the next paragraph, and it is not included in the other versions.

calculus will do; for example, a calculus to distribute nuts. (You invent a calculus good for practical purposes). But normally you would mean: this will come out as the other does.

If the properties of the nuts change entirely—if they split—you could say this calculus won't do any longer for practical purposes; and yet you will certainly say that it will do. You have a criterion here in your calculation. When you say the figures will recur, obviously you are not in the end saying that this calculus will help you to distribute nuts.—Of course your criterion is in a sense a practical one. That is why you have a proof. "It will do" is a preliminary to the proof. We won't say it if later on we find a proof that it's not correct.—Finding a proof does have to do with this always *coming right.*

Now suppose that you want to show a man that the figures must recur. You show him that the first remainder comes again. Is it easier for him to see that the figures will recur if we work through several cycles than if we've done only one? In a sense, *yes.* So you may present it to a child in this way.

"But isn't the mathematical proof just as conclusive when it stops here?" *Yes.* Or where does it begin to be conclusive evidence?

Turing: Proofs in mathematics vary from rather formal proofs to this kind of thing, where one just tries to show him how the thing happens, and if he is convinced then it is all right.

Wittgenstein: Yes. But first of all, one can easily put in steps so as to make this sort of thing more formal.

And secondly, what do you *convince* him of? He is convinced that it will go on thus. But what is that? Is he convinced that people will always go on in this way? That would be absurd. —And is he rightly or wrongly convinced?

But must you say that you are convincing him of anything? Isn't this a bad expression?—Think of intuitionism and saying, "We intuit that 3 follows 2." What is wrong with this is like *convincing* someone that 103 follows 102.

Wisdom: Instead of saying that he is convinced, couldn't you say that he accepts it?

Wittgenstein: Yes, that gets rid of one absurdity. You can say that you convince him that this is the natural way to go on. Though what you make him *accept*—is *this* as the rule of division.

Turing: The fact that the mathematical proof that the digits will recur is quite conclusive, even if it only goes to the point where the first remainder recurs, shows only that one has a technique for explaining such things to trained mathematicians which is different from that for children. It is not a question of the going to the point where the first remainder recurs being a mathematical proof as opposed to something else.

Wittgenstein: Yes.—And we might make a rule whereby in mathematical proofs one went past the point where the first remainder recurs and always went through ten cycles.

How do we convince someone that certain recurrences occur, without dividing? We say, "You see, don't you, that we must always get the recurrence."—Suppose I draw a rectangle that I divide into equal parts. I draw triangles here, and then I say, "You see, don't you, that I could go on drawing triangles until I had filled the whole rectangle—that if I draw enough triangles I get down to here."—Now how do you see this? You haven't tried it. You haven't even imagined it. You just draw the first two or three and then say, "If we go on doing it, we shall get down there."

"Well, must you in doing it get down there?"—Doing *what?* If you mean just "putting triangles together", I needn't get there at all. But then I'd say I hadn't done it right.—If I say this, I've already adopted it as a criterion for doing it in the right way, *that* I get down here.

Suppose I say, "If I put bricks on top of one another, I must get something which goes up straight:

But in fact I try, and I get something which goes, not thus, but so:

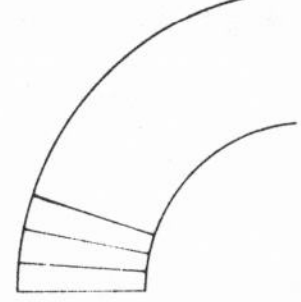

I should then say that I hadn't put them one *on* another, or the bricks weren't really square, or something of the sort. The point is: by saying they *must* go up straight, I am saying, "I have adopted it as a new criterion for their being square, *that* they go like this." I've decided to say, "No, something must have been wrong." You might say, "Out of a hypothesis, I've made a postulate."

It is just the same with 1 : 7 = 0.142857142 . . . You say, "This must give so-and-so."

Suppose it doesn't.

Suppose *what* doesn't?

Here I am adopting a new criterion for seeing whether I divide this properly—and that is what is marked by the word "must". But it is a criterion which I need not have adopted.—For just as bricks measured with all exactness might give a curve ('space is curved'), so 1 : 7 = 0.---- looked through with all exactness might give something else. But it hardly ever does. And now I've made up a new criterion for the correctness of the division. —And I have made it up because it has always worked. If different people got different things, I'd have adopted something different.

We could even say, "Bricks which are *really* square *must* give a curve. Space is curved." ("We're getting tired if we don't put in a 1.")

"We first believed that drawing the triangles would always give the rectangle, and then *proved* that it must."—The result of the proof is that we have persuaded someone to accept this as a new criterion.

He believed that if we follow the rules as we do follow them, being prepared as we are, then this is what will always happen. Then later he takes recurrence as the criterion: "It must happen." (Timeless 'must'.)

Put differently: Let's regard division as a geometrical drawing, or a pattern of dashes. We could say just that: if you put the same pattern below, you'll be able to draw a line here. Just like: "If you put this pattern on this one you'll get: ." The question of recurrence is then a strictly geometrical question: the man will be persuaded that if he repeats this pattern here, there must be the same numeral repeated. (A new criterion *that he has done* so-and-so.)—When you do the new kind of calculus, which is that with periods, you are now making an entirely new use of this pattern.

Today I did not at all get to the place I wanted to get to.

XIV

There are two points which hang together:

(1) The use of counting in mathematics, as contrasted with the use of counting outside mathematics. This was to lead up to the point that: with what right you call something a numeral can't be understood simply by looking at its use in mathematics.—No; that last remark is awful. But if you look at mathematics to see what justifies calling something a numeral, you are likely to make a terrible mistake. What its real use is, is in a peculiar way hidden. —But I will not talk about this immediately but about a closely connected point which came up the other day.

(2) The function of a mathematical proof—for example, a proof that a certain period always must recur.

Now Turing said that the things we call proofs in mathematics ranged from formal proofs to proofs which are mainly designed to convince you of something. This is sometimes considered a most important distinction. I believe that at the basis of this there is a huge confusion.

Suppose I say, "Some proofs are strictly logical—watertight, airtight, foolproof; and some proofs are meant merely to convince." It seems almost as though what is meant is that they should have a certain psychological effect.[1]

There are proofs of this kind. For example, in engineering you're taught a formula—you have a piece of iron which you twist, the formula is to give what moment is needed to do so-and-so—and the formula is proved. Three formulae are proved from three different assumptions. For the student, the formulae by themselves are quite enough. If you are studying engineering and I want to convince you, I could tell you simply, "This has been found by experiment." But the proof *satisfies* the student. (The proof might be used by some as a help to memory.) And one might call this a proof serving to convince you. But this isn't a proof of a mathematical proposition but a proof of physics.

But in mathematics if we talk of a proof serving to convince you, this is very queer. We may ask, "To convince you *of what?*" —Suppose we have a proposition in mathematics for which there are two proofs: (a) the real watertight proof, (b) the proof for the normal student, to convince him it is all right.—But: (1) the opposite of this proposition could be assumed and we could have a different mathematics; (2) it could have a different proof. It could be assumed without a proof at all.

Suppose that Gasking cannot see outside, and I want to convince him that it is raining. I might get his best friend, whom he always trusts, to tell him so. Here Gasking is convinced of something that can be got at in a different way. The criterion for raining is not normally what I have taken in this case—that someone says it is.

But in mathematics it is quite different. What is the criterion that the mathematical proposition is true? It's not the psychological proof. Is it the watertight proof or what?

Is it like this: God sees it is true. We can get at it in different ways. Some of us are easily persuaded that it is so, others need a long elaborate proof. But it is so.

But what is the criterion for its being so—if not the proof?

1. Cf. Hardy, "Mathematical Proof", pp. 16–19.

The idea seems to be that we get at the truth which was always there apart from the proof. (The proof is a kind of telescope.)

We call a certain structure a logical proof, say in Russell's symbols. Suppose I say, "This is the proof." Well, if it is, then we might say that we can't see the mathematical proposition in any other way than by seeing the proof.

```
   126 X 631
   _________
   - - -
     - - -
       - - -
   _________
= - - - - -
```

This process is a proof. "Could I have proved this differently?" Well, could I have got this formula by other means? Of course. But need I have proved it at all? Couldn't I have taken it as a primitive proposition?

Suppose someone says, "The proof serves to convince me that this is really so." In a sense this obviously is so. But let's see.

The proof was taken to be the multiplication. But could this be said under all circumstances to convince me? Would it if it were the only multiplication? Suppose somebody had written this down in a country where they didn't have multiplication. What the hell would it convince them *of?* We say it convinces us in a particular system. And in this case we call convincing, multiplying.

If you say it convinces you—it would be far more useful to say that outside the system of proofs, this makes no sense whatever. It would if anything be a perfectly useless rule of substitution, saying that instead of these signs I could write some others. This isn't what we now mean by this mathematical proposition, which has its sense *in* the system of proofs.

Suppose I could get at the same result by a chain of Russellian transformations. Should we say, "The multiplication is only to convince us, whereas the other is the real proof that it is so"? But what is it that is so if not *this:*

```
  126 X 631
  _________
  ---
   ---
    ---
  _________
= -----
```

If this is the technique of multiplying, there cannot be a *better* proof than this. Although there might be another proof that gives us new insight—by connecting this calculus with other calculi.

In this "convince" talk there is the constant muddle between mathematical and non-mathematical propositions. For the word "convince" is taken from the case where there is a direct test or criterion for something, and also more or less indirect ways of convincing someone.

Turing: When I used that phrase, I merely meant that a trained mathematician has prejudices in favour of a certain kind of proof.

Wittgenstein: Let's consider this. There are cases of mathematicians having prejudices. For instance, nowadays proofs by means of drawing lines are considered rather fishy, as opposed to proofs by writing, although they are just as good proofs. Why is there this prejudice?

Suppose I drew the proof of the construction of the pentagon. There easily arises a misunderstanding because on the one hand the drawing I make is no proof at all.—If a man knows no mathematics it convinces him that if you do certain things you get certain things: a figure with five sides that look equal, or one that is regular when measured. This is not mathematics, not a proof: it is a device for getting a particular appearance. It is neither a rigorous nor a non-rigorous mathematical proof.—But suppose I had shown you a construction of a pentagon here. Suppose I then said, "Now if you want to get a pentagon anywhere else, just copy this." Is this mathematics? I would say yes. Though not [if it were] the construction of [a measurable] pentagon. It says you get the same if the projection is done in a certain way.

The figure of the Euclidean proof as used in mathematics is just as rigorous as writing—because it has nothing to do with whether it is drawn well or badly. The main difference between a proof by drawing lines and a proof in writing is that it doesn't matter how you draw lines, or whether the *r*'s and *l*'s and *m*'s and *e*'s are written well.

[*Referring to a sketched figure*] This is perfectly all right.—It really is a prejudice that these figures are less rigorous; partly because the role of such a figure is mixed up with the construction of a measurable pentagon—mixing up drawing used as symbolism with drawing as producing a certain visual effect. (It is true that I may draw so badly you can't see what it's about, but then in writing, I may write so badly you can't read it. This is analogous to the case of a mathematician who doesn't know what an ellipse looks like and has to ask his son. He misses the symbolism if he doesn't know what ellipses look like. As if he didn't know what *r*'s and *l*'s look like—still more sublime.) This is one reason for the prejudice against drawing; another is that there is less opportunity for generality.

But let's get back to the question of mathematical proofs convincing us of something.

You might say, "Wittgenstein, I know what you're going to say. A proof doesn't convince us of something, but persuades us to adopt a rule." But that will not do. For how does it persuade us?—You might say, "Strict proof persuades us rightly, whereas other proof does so as a friend does."

But now about "A proof persuades us to adopt something as a rule"—there are all sorts of ways of persuading people. You could persuade a man to accept the pentagon construction, that is, to do so-and-so, by showing him what you get (namely pentagons that are regular when measured). This is not mathematical proof at all.

A mathematical proof persuades us by making certain connexions. It puts this (126 × 631 = – – – – –) in the middle of a huge system—it gives it a place.[2] We are taught to adopt any rule that can be produced in such a way.

If you say, "It persuades us to adopt a rule"—it looks as if there was a government which had various means, some more crooked than others, for persuading its subjects to adopt certain rules.

But what the proof does is to make the connexion: by this connexion it may or may *not* persuade you. It may in fact persuade you of the opposite. We may have a particular prejudice against these numbers or what not.

Consider Wisdom's case: a dispute he had once with his mathematical master at school. The master had said that $3 \times 0 = 0$ and Wisdom had said that $3 \times 0 = 3$. What would be said by each side to prove his case?

Well, Wisdom says that if three cows are multiplied by 0 then that means they have not multiplied at all, and so there are still three cows. The master might then say, "Look: $3 \times 2 = 3 + 3$; $3 \times 1 = 3$; and so $3 \times 0 = 0$." He makes Wisdom surprised at having to admit that $3 \times 1 = 3 \times 0$.

Suppose that one learns a system of multiplication which starts with 1 and then later introduces 0. When one first introduces 3×0, one has complete freedom to say either $3 \times 0 = 0$ or $3 \times 0 = 3$. In deciding which to say, we go by analogy; but everything is analogous to everything else.—How would you convince someone that $3 \times 0 = 0$? By getting him to use this technique; by showing him that it is the most useful one or the one used by most people. But this showing him that it is what most people say, or tempting him to follow the technique, wouldn't be called a *proof* that $3 \times 0 = 0$.

You might ask what the difference is between proving that $3 \times 0 = 0$ and proving that $126 \times 631 =$ - - - - -. Well, one is taught a technique which applies easily to things which are obviously not exceptions; but 0 is clearly an exception in one way or another.

2. Cf. *Bemerkungen über die Grundlagen der Mathematik* (Frankfurt, 1974), p. 303: "Die Beweise ordnen die Sätze. Sie geben ihnen Zusammenhang."

But let's get back to the main point. A mathematical proof connects a proposition with a system. But some connexions have greater permanence or greater force than others. You might even start by showing one connexion and then show another which seems more important than the first and makes you give it up. You may start by proving something and then go on to prove its opposite. In that case you're changing your mathematics.

Let's try to put this together with what we said about *believing* a mathematical proposition before it is proved. What are the criteria that we believe something? Take a particular theory of Eddington's about the end of the world: in 10^{10} years, the world will shrink or expand or something. He might be said to *believe* this. How does he do this? Well, he says that he believes it, he has arrived at it in a certain way, is rather pleased that he has reached this knowledge, and so on. But what could be called actions in accordance with his belief? Does he begin to make preparations? I suppose not.—Compare believing something in physics and the case where someone shouts "Fire!" My saying "I believe" will have different properties and different consequences, or perhaps none.

But how does one believe a mathematical proposition—say, that a period will recur in the division of 1 by 7—before one has a proof? Is it that one says with a feeling of conviction, "I am sure that it will always be so" or "I am sure a proof will be found"—or that I try to *find* a proof myself? This would certainly be an action in accordance with my belief. One of the criteria is certainly that I'm extremely anxious to try to prove it—as I might be anxious to prove Goldbach's theorem if I believed it.

Think of: "I believe that this proposition will be proved within the next twelve months."

What would be such a thing as a *hunch* that this will come out all right? A hunch for certain propositions in the theory of numbers, say.—At first sight we might say, "What could this mean?" But it may come out right if I write down several propositions and these are all proved in the next decade.

We might ask, "How can you have a hunch about these things,

since as long as the proof isn't there, it has either no meaning at all, or not the meaning which it will have when the proof of it is there?" One might put it: a mathematical proposition only gets its meaning from the calculus in which it is embedded. The use which we can make of this mathematical rule depends entirely on the mathematical system in which it is embedded. If we *only* had the rule $25 \times 25 = 625$, this would be nothing; it would be nothing we could do anything with.

Suppose someone had a hunch that "every even number greater than 6 is the sum of two primes". If you have a hunch it will come out right, you have a hunch that the mathematical system will be extended in this way—that is, that it will be best or most natural to extend the system in such a way that *this* will be said to be right.

Suppose someone said, "What you, Wittgenstein, say comes to saying we could *also* extend arithmetic in such a way as to prove this is not so, or to make it a primitive proposition." I'd say: certainly.

Because of course you haven't yet made this extension. The road is not yet actually built. You could if you wished assume it isn't so. You would get into an awful mess.

What you believe is that the system will be extended in such a way that it *will* come right.

You might say, "Wittgenstein, this is bosh. For if the system will be extended in such a way, it must be *capable* of being extended in such a way." If this is so, then the person who has a hunch that Goldbach's theorem is correct has a hunch about the possibilities of extension of the present system—that is, he believes something about the essence, the nature, of the system, something mathematical about it.

[*To Turing*] Would you be inclined to say something like that?

Turing: Surely when Wisdom had a hunch that $3 \times 0 = 3$, he did not mean simply that one could say that $3 \times 0 = 3$.

Wittgenstein: No, of course he doesn't. One might say he means that it will turn out that $3 \times 0 = 3$ is the only natural or most natural thing to say. But most natural for what, or under what circumstances?

What would it mean to have a hunch that $1 \times 0 = 0$ if this were taken as a primitive proposition? Taking it as a primitive proposition is just deciding on it as a rule.—If one has a hunch that a certain thing will come true, as opposed to a hunch that it will be *postulated* true (as in fact it may even be postulated false), then one's hunch is that it will be the most natural thing to say.

This doesn't satisfy. "If someone has a hunch that it will come true, then in order that it should be *able* to be true, it must reflect on the nature of the present system of mathematics."

If we adopt the idea that you can continue the road either in this way *or* in that way (Goldbach's theorem true or not true) —then a hunch that it will be proved true is a hunch that people will find it *the* only way of proceeding. Though before anyone has found a proof we could say, "If someone has found a proof we have a perfect right not to acknowledge it."

Can one find a contradiction in a certain system?—One might say, "It depends on you." One might say, "Finding a contradiction in a system, like finding a germ in an otherwise healthy body, shows that the whole system or body is diseased."—Not at all. The contradiction does not even falsify anything. Let it lie. Do not go there.

[*To Turing*] Your tendency is always to avoid one simile.

If you say, "The mere fact that a proof *could* be found is a fact about the mathematical world", you're comparing the mathematician to a man who has found out something about a realm of entities, the physics of mathematical entities. If you say, "You can go this way or that way", you say there is no physics about mathematics.[3]

Professor Hardy says, "Goldbach's theorem is either true or false."—We simply say the road hasn't been built yet. At present you have the right to say either; you have a right to *postulate* that it's true or that it's false.—If you look at it this way, the whole idea of mathematics as the physics of the mathematical entities breaks down. For which road you build is not determined by the

3. (From "Can one".) The organization of this passage rests on conjecture. The paragraph on contradiction is only in B; it is not clear where it belongs. It may possibly have been addressed to Turing. The rest of the passage, which is based on B, S, and R, also involves guesswork.

physics of mathematical entities but by totally different considerations.

The mathematical proposition says: The road goes there. Why we should build a certain road isn't because mathematics says that the road goes there—because the road isn't built until mathematics says it goes there. What determines it is partly practical considerations and partly analogies in the present system of mathematics.

But the fact that a proof of the theorem is *possible* may seem to be a mathematical fact—not a fact of convenience etc.

What would you say against the person who said, "Goldbach's theorem can be assumed true or assumed false; therefore there is no question of looking to see which it is"?—Professor Hardy says: Goldbach's theorem is true or false depending upon the mathematical facts; it is not a matter of rules or convenience; it is a theorem concerning reality.[4] How would you defend this?

I said (in reply to Hardy's idea of a world of mathematical entities that the mathematician looks into) that the mathematician is a man who builds new roads, or invents new ways of thinking. I tried to tempt Turing to say that the fact that a certain extension of the mathematical system is possible is a *mathematical fact*, the idea being that once the proof is given, doesn't this show something about reality?—But unfortunately I failed to tempt him.

We can easily persuade a normal person that the right hand and the left hand (you can take two more mathematical things if you like) cannot be superimposed. Wouldn't one be tempted to say that it is a mathematical fact: that if we prove this, we have proved a mathematical fact?

When Hardy says, "Goldbach's theorem is either true or false", he means "The numbers either *have* this structure or they *haven't* this structure." Similarly we might say, "These two structures have the quality that they can't be superimposed."

You could of course *show* him that they can't be. Couldn't he

4. "Mathematical Proof", p. 4.

say, "Yes—but this only means that you haven't yet succeeded"? Suppose he went on trying, and said, "There is a position I haven't been in. How do you know I've been in all possible positions?"

The normal person doesn't do that. But couldn't you imagine a fixed prejudice, due perhaps to the fact that they are, as we say, identical, that they *must* be capable of being superimposed? Here you have two ways we might go. There is an extension, there is a road which runs that way too. For how is he to know he has tried all possible positions?—We could even agree with him if he gave up trying to do it and behaved as we do. If he went on trying it would be disagreeable—although he might call it a mathematical pursuit.

If you say, "Mathematical propositions say something about a mathematical reality"—which expresses a natural tendency—a result of that tendency would be roughly this: We say certain things about animals. There are propositions we all know, and propositions about exotic animals, which have a certain charm. If you have the idea that mathematics treats of mathematical entities, then just as some members of the animal world are exotic, there would be a realm of mathematical entities that were particularly exotic—and therefore particularly charming. Transfinite numbers, for example; as Hilbert says, this realm is a paradise.

I'm not going to say, "If we look into this, we'll see there are no transfinite numbers"—that is all bosh. But I'll say: "What, for example, a transfinite number is the number *of,* does not appear at first sight if you look into mathematics."

The number 3 can be the number of apples or the number of roots of a certain equation. If we didn't know how to use 3 *outside* mathematics at all, we should get no idea of its use if we said it is the number of roots of this equation. For in mathematics 3 is the number of roots either by definition or by proof.

"Having the number so-and-so" is used differently *in* mathematics and *outside* mathematics. So with the expression "having the number $\aleph_0$"; it will be wrong to think we know how it is used if we say there are $\aleph_0$ cardinal numbers—which have it either by definition or by proof.

"This child knows $\aleph_0$ multiplications."—Do I mean to say that therefore $\aleph_0$ is finite? Not at all, I say only that in order to find out the role a number plays, you mustn't look to repeating a mathematical phrase like "There are so many so-and-so's."

If you say that mathematical propositions are about a mathematical reality—although this is quite vague, it has very definite consequences. And if you deny it, there are also queer consequences—for example, one may be led to finitism. Both would be quite wrong. There is a *muddle* at present, an unclarity. But this doesn't mean that certain mathematical propositions are *wrong,* but that we think their interest lies in something in which it does not lie. I am *not* saying transfinite propositions are *false,* but that the wrong pictures go with them. And when you see this the result may be that you lose your interest. It may have enormous consequences but not mathematical consequences, not the consequences which the finitists expect.

Similarly, some people have objected to the differential calculus: "It can't treat of the infinitely small; for one doesn't find anything infinitely small in it." This is queer—we might ask, "What would it be like to find anything infinitely small in it?" You might ask, "Why are you disappointed now?"

Yet they were getting at something important. For one might say that there isn't even anything *small* mentioned in the calculus, since the calculus is quite different from its application. There is no mention of size.—Is it a consequence that we should drop the idea of the infinitely small? Not at all. All we can say is that the expression "infinitely small" is misleading because it gives the wrong picture, makes one think of very tiny things.

Suppose I said, "The child of eleven has learnt $\aleph_0$ multiplications."

"Well, what's infinite about it?" Well, $\aleph_0$ is infinite about it. That's all. But to say "There's something infinite about it" suggests "There's something *huge* about it." What is huge about

$\aleph_0$? The child who has learnt $\aleph_0$ multiplications hasn't learnt anything huge.

Does this show that there is no such thing as infinity? Not at all.—If I say it's misleading to use 'infinite' here, this does not interfere with the mathematics of the matter.

"I bought something infinite and carried it home." You might say, "Good lord! How did you manage to carry it?"—A ruler with an infinite radius of curvature.

You might ask, "Is there anything infinite about a small thing like a ruler?" But why should we not say, "Yes, the radius of curvature"? But it's not huge at all.

I'm really trying only to examine the difference between counting in mathematics and ordinary counting, and the difference between a mathematical proposition and an experiential one.

XV

It has been said very often that mathematics is a game, to be compared with chess. In a sense this is obviously false—it is not a game in the ordinary sense. In a sense it is obviously true—there is some similarity. The thing to do is not to take sides, but to investigate. It is sometimes useful to compare mathematics to a game and sometimes misleading.

There is an argument used again and again *against* the idea that mathematics is a game: "All right—it is comparable to chess. Moves on the board can be compared with transformations of mathematical expressions. But in chess we must distinguish between (1) games played by different people; (2) the theory of the game." [1]

If you compare mathematics to a game, one reason is that you

1. Cf. Frege, *Grundgesetze der Arithmetik,* Vol. II, § 93, pp. 101–102.

want to show that in some sense it is arbitrary—which is certainly misleading and very dangerous in a way.—Now I have said things which could be interpreted in this way: "You could do it another way", etc.—But if you say the rules of chess are arbitrary, your opponents will say the *theory* of chess is not arbitrary. If you prove that you can't mate with two knights, that is a fact, a truth—and is not arbitrary.—So if you had part of mathematics which was a game, then anyway there would be another part—the theory of the game—which would not be a game and would not be arbitrary.

Now in what sense are the rules of chess arbitrary? Suppose I said, "The colour of a pair of trousers is arbitrary." As far as strength goes this might be; it certainly has no influence on that. But grass green trousers would hardly sell.

You might say, "Another rule would have done just as well." For what? This suggests that there is nothing in the object of the game which determined this rule. Well—

We don't make up the rules of these games. Chess and similar games we have *inherited.*

To a man who invented chess, everything in it may have been very important—no more arbitrary than a poem is arbitrary. ("Not a comma to be changed.") [2]—We may say that the rules are arbitrary in the sense that I wouldn't play it like this if I hadn't learned it like this.

Couldn't the game of chess be much less arbitrary still? Of course. It might be an exact picture of our battles, say. We could imagine that an arrangement of chess men might be a picture of the movements of warriors.

Suppose chess were used in this way: We know a tribe which sometimes decides certain very important questions—whether there shall be a war or not—by doing what we should call playing a game of chess. Two parties sit and make certain moves. Suppose it were very important for us to discover what will happen. We might then learn chess only in order to learn what their

2. Cf. *Philosophical Investigations,* Part II, § xii.

decisions will be, what they certainly will not arrive at, etc., etc. —Here actually we could use the game only in order to calculate things. It might not be a game at all. It might be done merely in the Foreign Office; there would be no chess clubs.

Now what is the role which "the theory of the game" plays in relation to the game? In the theory of the game certain propositions will be proved: for example, "You can't mate with so-and-so", "You can't mate in so many moves from such-and-such positions", etc. Obviously if you have proved this, you know something about the game—about what will happen. Now should we say that this fact—that you can't mate with two bishops—rests on mathematical facts?

Should we say it rests on facts of physics (about the rigidity of the chess board, etc.), or on psychological facts, or should we say it rests on certain mathematical facts? There is an enormous temptation to say the latter.

The other day I tried to put the problem this way: If we have a question like "Will Goldbach's theorem be proved?" one may answer, "You can assume it as true or as false; take this or that as a proof of it." Here you get something like "It's arbitrary." Against this you get: Suppose someone produced a proof. Couldn't one say that the *possibility* of this proof was a fact in the realms of mathematical reality? In order that he should find it, it must in some sense be there. It must be a possible structure.

In the case of chess the same point comes up. Doesn't this "We can't mate with two bishops" rest on mathematical fact? We might say: It is a question of mathematical possibility. The question is "Is it possible?"—not whether anyone will ever try it. Isn't there such a thing as mathematical possibility?

Frege, who was a great thinker, said that although it is said in Euclid that a straight line *can* be drawn between any two points, in fact the line already exists even if no one has drawn it. The idea is that there is a realm of geometry in which the geometrical entities exist. What in the ordinary world we call a possibility is in the geometrical world a reality. In Euclidean heaven two points are already connected. This is a most important idea: the idea of possibility as a different kind of reality; and we might call it a shadow of reality.

We multiply 25×25 and get 625. But in the mathematical realm 25×25 is *already* 625.—The immediate [objection] is: then it's also 624, or 623, or any damn thing—for any mathematical system you like.—If there is a line drawn there between two points, there are 1000 lines between the points—because in a different geometry it would be different.

I'm not saying anything against that picture. We don't yet know how to apply it. There would be an infinity of shadowy worlds. Then the whole utility of this breaks down, because we don't know which of them we're talking about. As long as there is only one, we know where to go to find out what we want. I am to make an expedition into one—but which? You might say, "I want to go into a world where a straight line really does connect two points."—Yes, but there is an infinity of those. And an infinity of consequences follow, etc.

You never get beyond what you've decided yourself; you can always go on in innumerable different ways. The whole thing crumbles because you are always making the assumption that once you are in the right world you'll find out.—You want to make an investigation, but no investigation will do, because there is always freedom to go into another world.

This doesn't at all destroy Frege's argument; it merely shows there is something fishy.

"What for us is possibility is reality in another world. In another world things are drawn extremely faintly—the geometrical line is very faint and thin—and we *trace* it:  . We draw a more or less rough line over it."—Now there are cases where this picture applies, where we treat this as the criterion of possibility. But let us go back a bit.

We imagine possible structures and impossible structures, and we distinguish both from real structures. It seems as though in mathematics we showed what structures are conceivable, imaginable, not what are real. We'll prove that there *can* be such-and-such a construction. What does this mean?

"Imaginable" is a term which is very good in some ways and in some ways very misleading. "Can imagine" or "can't imagine" have a psychological sound. "Can you imagine Lewy without a head?"—This depends on the vividness of your imagination. —On the other hand [what we call possible] has something to do with imagining.

If you want to understand what logical possibility is, one analogy is chemical possibility. Take a structural formula in chemistry:

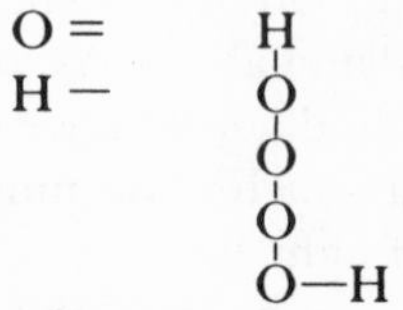

```
O =        H
H −        |
           O
           |
           O        H2O4
           |
           O
           |
           O—H
```

"H_2O_4 is chemically possible although it doesn't exist." Now what is it that is possible about H_2O_4 if it doesn't exist? Is it a bit more *real* then by being possible? Is it nearer to existence than HO_2? You might say, "Are there stages of existence?" (Compare: "It didn't happen but it was quite possible, which is something.")

What is the use of this idea of its being a chemical possibility? Is there some shadowy reality here too? And where is the shadow?—Couldn't chemists think there is a world in which there is some H_2O_4, although it hasn't come into reality?

Isn't there a shadowy reality here—that you have actually drawn? Isn't your *language* the shadowy reality?

When you say "H_2O_4 is possible" you simply mean it is a sign in your system. That system of valencies wasn't chosen at random, but because it fitted well with the facts. But once chosen, what is possible is what there is a picture of in the valency-language. We have adopted a language in which it *makes sense* to say "H_2O_4 . . ."—it isn't true, but it makes sense.

Just the same is true of geometrical possibility. To say that a straight line can be drawn is to say that it makes sense to talk of a straight line being drawn.

(The idea of a straight line as thin and shadowy comes from

our being accustomed to using dashes for lines and dots for points.[3] If we painted sheets of paper and used the colour boundaries as lines, no one would say that lines are thin; for we do not call colour boundaries thin. Nor would they say that we cannot imagine lines, which is all bosh because the edge of a dash is a geometrical line.)

Let's go back a little. Does the fact that one cannot mate with two pawns "rest on a mathematical fact" or ". . . on a mathematical reality"? One might say, "God when he created things made it *possible* to mate with so-and-so. He created the world with certain mathematical properties, through which this is *impossible*."

"God made it impossible in such a game as chess to mate with two pawns": if that can be said, then it can also be said that this is the mathematical world he did create.

Gasking cannot mate me with two pawns (if he doesn't cheat, of course). We could ask: What is it he can't do? What situation is it he is incapable of bringing about? What is it *like* to mate with two pawns? What does mating consist in?

Suppose we say, "That he can't mate with two pawns—rests on the mathematical fact that a certain structure is impossible"—now what do you know about mating? How is 'mating' to be defined? Either the definition might include the impossibility—or possibility—of mating with two pawns. Or mating might be defined by analogy: "It is this sort of thing . . ."—and then how should we know? [4]

(a) "It's a mathematical fact that he can't mate with two pawns."

(b) "It's a mathematical reality that he can't mate with two pawns."

Could we say (a)? Yes. It means it's a mathematical proposition that he can't . . . —What would (b) suggest?

3. This sentence is inaccurate in B and incomplete in S; the version given is a guess based on both.

4. (From "What is it *like*".) The four versions of this passage were quite different and could have been put together in other ways.

When I say (a) I don't have a picture in my mind that it corresponds to a particular part of reality where these things are found; I'm not comparing it with zoology or botany.—I'm only pointing out a particular *use* of this proposition. It is in fact expressed in this way: If he *seemed* to mate with two pawns, we'd say, "Somehow he cheated" or "We've overlooked something."

But couldn't we say that a reality does correspond to it? Yes. I can be absolutely sure that it will never happen. Now what won't happen?

What is it we are sure of: If a man has two pawns and a king, we give up playing. Why doesn't he keep on trying?—I can prove the thing and satisfy you: that there is nothing I will show you here that you will ever call "mating with two pawns".

But a man may not give up trying—like the man who went on trying to make the right and left hands coincide. He may hold that there are still possibilities which we have not taken account of.

Further, is it inconceivable that he should one day do something which he and everyone else would call mating? No, not at all. He might do something which we *now* should call "not playing the game" but of which people then would say "why yes, that's all right".—It seems to me immensely unlikely and I'm not going to gamble on it, but it's conceivable.

We might ask, "What does an impossible structure look like? Can it be described?" When we call anything "impossible" in a logical sense, we call nothing its description, we eliminate its description.

"You look in the shadow world."—You look in a book.

This: H is the shadow world.

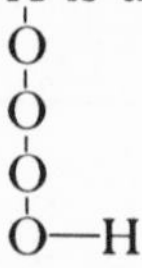

The point is you don't *look* in the shadow world, you *construct* things.—Or it means: you learn chemistry.

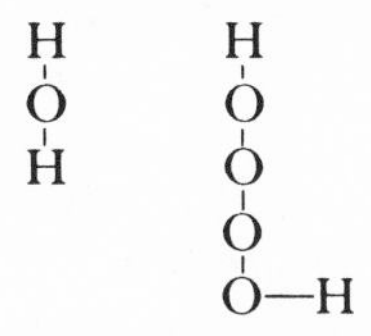

"The one has something corresponding to it. The other has no counterpart in the real world."—But what is its counterpart in the shadow world? What stands for the number 3 in the shadow world? Not | | |, or three apples. You can't think.—You could imagine a realm of spider-web lines, etc., a space somewhere in heaven reserved for Euclidean geometry. All points would be connected by straight lines—but would all straight lines be bisected? All constructions be done? But you could imagine the shadow world to consist simply in a copy of Euclid's *Elements.* There is no need to project the thing into a universe of its own.

It was said that the rules of chess are arbitrary, whereas the theory of chess is not. Here there are facts, and they seem to be mathematical facts.—I should say there are all sorts of important facts about chess. We play it in such-and-such a way. If you move a piece from one square to another, it hardly ever happens that you don't know, for example, whether you've moved it one square or two. We all agree as to where the pawn was, how it moved, who had the black pieces, etc. It happens that we forget but not very often.—If these things happened more often, if a player often had a lapse of memory and didn't know—we couldn't even *say* "We all have lapses of memory." The question couldn't even arise as to where it was.

In the heat air rises from the pavement and shimmers. Sometimes we see shapes shimmering through it, and eddies in it. And if I asked you what shapes you saw, you couldn't describe or draw them exactly. In certain cases, you'd have to say, "Oh, I just see shapes." It would be more like this in the case which we now call constant lapse of memory. In less extreme cases, he would say,

"The pawn was somewhere here", "I suppose I can now take his knight", etc.

These are important facts about the game.

Could one now say what role the mathematical theory of chess plays in respect to chess; and in what sense the mathematical theory of chess is *not* arbitrary, while the rules of the game are arbitrary? How far is it correct and how far incorrect to say that the mathematical theory of chess expresses truths? Or could we say that the mathematical theory of chess is another game?

Turing: From the mathematical theory one can make predictions.

Wittgenstein: Yes, one can. But what sort of predictions? What is the relation between the mathematics of chess and the predictions?

In a sense the pure mathematics of chess makes no predictions. That is one of the important points. The pure mathematics of chess is like the pure mathematics of astronomy. The calculus makes no predictions, but by means of it you can make predictions.

"The difference between chess playing and the mathematical theory is that the one can be used to make predictions, but the other can't."—Chess playing as a matter of fact is not used for making predictions, but it *can* be—whereas the theory *is* used for making predictions; and that is the difference. They can both be called mathematics and both be called games. They could both be used for making predictions.

"The theory of chess is not arbitrary."—It's not arbitrary, mathematics is not arbitrary, only in this sense, that it has an *obvious* application. Whereas chess hasn't got an obvious application in that way. That's why it is a game.

The difficulty in looking at mathematics as we do is to make one particular section—to cut pure mathematics off from its application. It is particularly difficult to know where to make this cut because certain branches of mathematics have been developed in which the charm consists in the fact that pure mathematics looks as though it were applied mathematics—applied to itself. And so we have the business of a mathematical realm.

In order to see this, I'd like to consider an example: What would happen if we defined the number 2 as the number of roots of a quadratic equation? (This is not circular: we could write a quadratic equation without a 2.) I want to compare:

(1) "A quadratic equation has two roots."

(2) "There are two apples on this tray."

The first one seems a fact. The sentence has a structure like "A man has two eyes." Is there any essential difference in the way in which the proposition (1) is used and the way "A man has two eyes" is used? It seems as though we counted to get both.

In grammar there are four cases: nominative, genitive, dative, accusative. We could have two or six. We *distinguish* four cases. This is something quite different from *counting* four cases.

So we distinguish two roots in quadratic equations. But we could also distinguish four roots—always two and two alike. Would it make any difference to say a quadratic equation has four roots? We'd say it would introduce unnecessary complication. But it might not.—It isn't at all clear how we are to count the roots. We might count each root as double or we might count them both as one.

I don't mean—again—that it is arbitrary. There are very good reasons for saying there are two roots.

I shall now say something very absurd and then stop.

I can say, "I have a fork with two prongs. I've counted them." But suppose this figure were used as a paradigm for something——counting, say. Then if I said

"I count these (1 2) ",

it's clear that I could count them

(1,2 3,4) or (1,2,3 4,5,6) etc.

If I use it for counting, I may use it in all sorts of ways. The convention might be that there are two things to each prong; or that three people must take hold of each prong while I am counting. I might say then, "The fork has six prongs"—meaning that it is to be used in such-and-such a way.

I don't want to know in this case how many prongs there are, as I might in finding there are two apples in this room. If it is a question of coordinating numerals to them, I may do it in all sorts of ways.

In connexion with the roots of an equation you may be prejudiced by the fact that you always write down two things—two numerals. But whether these are to be counted as two *roots* (where 'root' is a mathematical symbol) is quite another matter.

There may be a good reason for counting one numeral as two roots or there may be no reason. But it would not mean that we should have to count chairs and apples like that. The mathematician would not count the expressions written down as four; he would not say Lewy had written four lines when we said that Lewy had written only two. But he would count the two expressions as four roots.

One might say this would change the meaning of the word "root"; it would change the way it was applied.

Next time I shall try to show the connexion between saying that the number of roots is 2, and saying that the number of cardinal numbers is $\aleph_0$. I shall also deal with the idea of a one-one correlation between cups and saucers and between odd and even numbers.

XVI

"A quadratic equation has two roots."

"A man has two eyes."

We are inclined to say, "We count in order to find the number of things." Compare: "We weigh in order to get the weight." Both statements are in a way equally fishy.

Counting is a method of producing a numeral. The interest in doing this lies in the application. We count in order to make predictions (whether there are enough chairs, etc.); we count for all sorts of reasons.

The mere saying of the words "one, two, three . . ." and pointing at objects—this might be anything: a game or a song or 'counting out'. I could equally say the words "Mary, had, a, little, lamb."—Why we call "one, two, three" and so on counting and *don't* call "Mary had a little lamb" counting is because of the application we *generally* make. "Mary . . ." in Chinese might be counting.

On the face of it, if counting is saying "one, two, three" etc., then counting the roots of an equation is counting. But if we say we count the roots—this doesn't necessarily mean that if we have a plus and a minus we count the signs here "one, two". We may count them "one, two, three, four".

If we said, "There are four roots here, always two and two equal", then what? There is no reason why we shouldn't correlate two numerals to each of the signs. What it would come to in application is that if I said to Lewy, "Write down two roots", he might write down the same expression twice; that is what we would call "writing down two roots".

"But then why shouldn't you count chairs in the same queer way, using two numerals for each?" You can, but this is a different case. If I want to count chairs to make a prediction, then I must use some technique of counting made up beforehand. It could be that chairs were counted in one way, students in another, apples in another, etc.

If you say, "Count the roots"—you might say this: "Suppose we had a technique of solving algebraically quadratic and cubic equations, and someone said 'Now count the roots of these equations.' Then the most natural thing would be to count so that a quadratic equation had *two* roots."

I might agree—but one couldn't even say, "It is obviously the most natural way to count 'one, two' ", since the roots may just differ in sign and may have the same number. It might in some circumstances be most natural to say, "Some quadratic equations have only one root, and some have two."

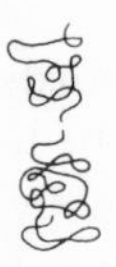

How many parts has this? You might say it has *two.* That might be the most natural thing to say—but couldn't you say *any* damn thing?

When I first said, "How many parts?" it's arbitrary; you don't know what to say. But if the upper part frequently occurred by itself in some context, and the lower part also—if they were, say, two Chinese letters—then we should say at once, "This has *two* parts."

Just the same applies to quadratic equations. If you *only* have the sign, it isn't clear offhand how you should count it:

$$-\frac{a}{2}+\sqrt{\frac{a^2}{4}-c} \qquad \text{(one or more??)}$$

Suppose we said, "This has two parts." Would this be absurd? Not at all. It would really be laying down a rule. If you say, "These are really two parts, only looking alike; so are these; and thus there are four"—this is all right, but the application will be different. And whether one rule or the other should be adopted depends on consequences, practical consequences. It might be that the reasons are overwhelming—as they are for counting $x = \pm y$ as two roots—or they may not be, or there may be different degrees.

Suppose that we learned to count, by counting, not things like chairs, but the parts of these: . The teacher as an exercise says, "How many parts are there here: ?" We say "one, two". "Now count these here: ." We say "one, two, three, four" (pointing) or "one, two, three, four, five".

Would he learn to count? The queer thing is, he would. He'd learn to say these words. But in another sense he wouldn't learn to count: he would not learn at all the application of the numerals, because he learns in a case where counting does not have the normal consequences.

If the proposition "A quadratic equation has two roots" *stood alone*, it would be as meaningless as "25 X 25 = 625" would if it stood alone outside any system of multiplications—although it is English and it looks all right. It would have no more sense than "This has two parts."

Instead of saying "two parts," we might learn to do this with brackets:

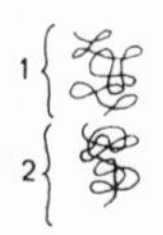

But what would you have learnt by that? Nothing. You just see a lot of lines, figures, etc. It would be of use only as part of a certain system.

The point is this. We learn a particular technique of coordinating numerals to structures which we call roots. We might learn to count the roots of a quintic equation, but we do not thereby know the mathematical proposition "The quintic equation has five roots."

Wisdom: If someone has been taught to count marbles, and is later taught to count apples, has he been taught a new technique?

Wittgenstein: Consider the case of people who count sheep and men—only. A child when he learns to count learns a series of words by heart and points at things, but he doesn't at first learn to use this as the men do. When he learns the practical application—to count sheep and men to see whether there are as many sheep as men—does he learn a new technique? Does he learn a new technique if he also learns to count nuts? We might say it was a new technique because it's nuts, not sheep; or say it is an old technique because it's so similar.

Compare the phrase "a new technique" with the question about "of the same kind", or "in the same way" and "in a different way": do I use the hammer in the same way when I use it to drive a nail and when I use it to knock a peg into a hole, etc., etc.

I want to get on to a terribly difficult business—a real morass—Russell's definition of number. It seems as though, if number is defined in this way (or Frege's way), everything will be clear. If we now know what *two* is, how can we be *doubtful* if there are two or four things here [roots]?

Turing: It is all quite clear. It is a fairly simply logical matter to show that there are two roots to a quadratic equation of a certain kind.

Wittgenstein: But can you see how this definition of number, which ought to have made everything clear, won't bring us any nearer?

The definition: "A number is a class of classes similar to a given class." 'Similar to' means (we will say for the moment) one to one correlated.

The difficulty recurs: what do we call one-one correlating the roots with so-and-so? We might call it anything.

You have substituted another expression for number.—This is the worst of all logical superstitions: the idea that if you write "There is an *x* such that" and the like, you can solve difficulties of this kind.

I don't want to run down Russell's definition. Although it does not do all of what it was supposed to do, it does *some* of it.

You've got to decide what you're going to call a one to one correlation with so-and-so.

It's as though we tried to see whether Lewy had two arms or one by drawing on the blackboard and saying, "One, two . Every man has two arms, so Lewy, being a man, must have two arms." Then if he has only one arm, we can count it as two; if no arm, we can point to each side of him in turn, saying, "There is one, two."—But the point is that we have only made transformations on the blackboard, and have not begun to get down to the question of finding out how many arms Lewy has.[1]

1. The three versions of this paragraph (B, R, and S) were quite different and there were a number of difficulties in putting them together.

"Similarity is decided by one-one correlation."—"Being correlat*ed* to"—or "being *able* to be correlated"? (Compare: the possibility of drawing lines.)—We might say that cups and saucers are correlated if every cup stands on a saucer, and otherwise they are not correlated—or they are correlated in a different way.

We must always think of *number* as we think of *length* or *weight*, and of *counting* or *correlating* as we think of *weighing* or *measuring*. We say that two things have the same weight if on the balance they counterbalance each other, or counterbalance the same number of weights.

If we are trying to find out whether there are the same number of people here as in the next room, one method we can adopt is one-one correlation: we could tie a string to each man here and attach it to one there. Then if there is one without attachment, . . . This is one way of finding numerical equality.

A totally different way is by counting—saying series of words.

A totally different way again is arranging things in geometrical figures.

. . . .
. . . .   Make a pattern of them, which
. . . .   we might call a counting-pattern.
. . . .

Must these different ways agree? Of course not. On the whole they *do* agree. We don't get into difficulties. But there is no more necessity that they should agree than that different ways of weighing should give the same result.—So we may have "the same number in different senses".

One way of finding numerical equality is applicable only to small numbers:  ; looking at them. We don't count or correlate—but decide straight away.

Correlating in real life (by string, positions, etc.) is to correlating in mathematics just as *joining* in real life is to joining by a straight line in geometry.

Suppose I said, "We will define all numbers by means of one-one correlation." Russell says that a class has this number if it can be correlated one-one to a standard class. This is like "A thing is a foot long if it can be made to coincide with the Greenwich foot." Is it always possible to bring it to the Greenwich foot?—But suppose nevertheless we defined:

"a thing is a foot long" = "it is possible for it to coincide with the Greenwich foot"

Suppose it is impossible to carry this out. Is the definition now absurd? Not at all. But it means now that the whole thing is reversed. For we have to say that if a thing is a foot long by other criteria, it could be made to coincide.

[Similarly with Russell's definition.] At first you thought of cases where the correlation was the criterion. But if the correlation isn't possible, then it is the other way round: if they have the same number by such-and-such a criterion, then it is possible for them to be correlated.

"It is possible for it to coincide"—doesn't it depend on what way? We might say: If we heat or stretch this, it concides with the Greenwich foot. Now this must either be forbidden or allowed.

Compare: The propositions of Euclid could be about length only if the conditions of measuring were laid down. But there aren't any laid down.—But they are some of the rules governing our use of the word "length".

If I have left out the way of making this coincide with the Greenwich foot, thereby I've left out everything.

If I just define: "The property of being a foot long is the property of being able to coincide with the Greenwich foot", then this so far doesn't determine at all what is a foot long, unless I have already fixed a method of comparison. And if we have many methods of comparison, the definition no longer enlightens us as to the meaning of 'being a foot long'. I may say, "If the apparatus yields this result, it is a foot long, and by definition it will be able to coincide." The criterion of its being capable of coincidence has disappeared. The definition seemed illuminating because we immediately think of one method of comparison.

Likewise with Russell's definition of number. One method of comparing numbers is one-one correlation.

Suppose Russell said: We only say that a class has ten members if each member can be one-one correlated with an intersection point of the pentagram. The pentagram is the standard 10 (so we can't say *it* has ten intersection points, any more than we can say of the Greenwich foot that it is a foot long). But here we might have all sorts of things which we can call correlating, and it is not clear which correlations should be taken as equivalent.

Russell gives us a calculus here. How this calculus of Russell's is to be *extended* you wouldn't know for your life, unless you had ordinary arithmetic in your bones. Russell doesn't even prove $10 \times 100 = 1000$.

What you're doing is constantly taking for granted a particular interpretation. You have mathematics and you have Russell; you think mathematics is all right, and Russell is all right—more so; but isn't this a put-up job? That you can correlate them in a way, is clear—not that one throws light on the other.

Russell puts down $(xy)(uv) \supset (xyuv)$ and proves this is a tautology. But suppose you had a *great* number of terms—ten million on each side—what would you do? You say you will have to correlate them. Here—$(xy)(uv) \supset (xyuv)$—it looks as if there were just one way of correlating. But with the huge number—would you correlate them in the same way?

Is there only one way of correlating them? If there are more, which is the logical way?—You can do any damn thing you please.

If you really wished to prove by Russell's calculus the addition of two big numbers, you would already have to know how to add, count, etc. And if you added differently, you'd get something different.

And the difficulties which crop up in the higher plane recur on the lower plane.

The calculation holds if a certain expression is tautological. But whether it *is* tautological presupposes a calculation.—In the case of a thousand terms it would be by no means obvious.

How could Russell's calculus decide how many roots an equation has? How would it enter? In this way:

Russell's calculus translates all the English you use in mathematics into symbolism as well. Thus "root" will be defined in Russell's symbolism. But in *what* way? What guides him in defining the word "root"? Russell replaces an English argument by an argument in Russellese, producing in his calculus a photograph of our normal usage. He could manage it in such a way that, *given* a certain definition of "root", it would be *extremely* natural to decide to count them in a certain way.—But just as we can use the word "root" in different ways, so he can correspondingly define it.

Suppose we are given *only* the calculation for quadratic and cubic equations. Then in English I introduce "root" and I say, "Here we have the roots." But how we are to count the roots is not yet laid down.—Russell can now give a definition of "root" in such a way that it will be most natural to count them in one way. But this does not mean that from Russell it *follows.*

We said in the first few weeks that we could, if we like, make 22 and not 21 come after 20. One might say, "Oh, but in Russell it is all quite clear and certain that 21 comes after 20." But how could it be clearer in Russell than it is in our ordinary language? That we know how to count is presupposed in Russell's definition of number.—He could have a definition such that it would be most natural to count in another way; but the definition doesn't *fix* the way of counting—or (similarly) what in this case we are going to call one-one correlation.[2]

A one-one correlation is nothing but a picture

And you can use this in all sorts of ways. About how it is to be used, Russell tells us nothing—except where he uses it himself. And he uses it in mathematics itself.

It is said to be a consequence of Russell's theory that there are as many even numbers as cardinal numbers, because to every cardinal number I can correlate an even number.

2. (From "But this".) None of the accounts of this passage was complete. B's was the fullest and made the general point clearer than did the others. Alterations were made in the material from the others where necessary.

But suppose I say, "Well, go on—correlate them." Is it at once clear what I mean? Is there only one technique for correlating cardinal and even numbers?

You can interpret "correlate" in such a way that you'll say, "Yes, there are as many . . ." But in what sense can you say you have *proved* this? You do a new thing and you *call* it "correlating them one-one"; and you call an entirely new thing "having the same number". All right. But you have not found two classes which have the same number; you have only invented a new way of looking at the thing.

In fact, if you say you have one-one correlated the even numbers to the cardinal numbers—you have shown us an interesting extension of this idea of one-one correlation. But you haven't even yet correlated any two things.

Supposing Lewy has learned to count. Then at a certain point we would say, "He can now count indefinitely." And we can then say, "He has now learned $\aleph_0$ numbers."

I wanted to describe today the relation between the actual *use* of the word "counting" outside mathematics and its use *inside* mathematics.

XVII

The question is: What is one to call a one-one correlation? You have the example of the cups and saucers; and then you think you know under all circumstances what the criterion of one-one correlation is. Russell doesn't bother about this, any more than Euclid bothers about fixing a method of measurement. This isn't a weakness or a strength.

Russell seems to have shown not only that you *can* correlate any two classes having the same number, but also that any two classes with the same number *are* correlated in this way. This at first sight seems surprising. But he gets over this, as Frege did, by the relation of identity.

There is one relation which holds between any two things, a and b, and between them only, namely the relation $x = a \,.\, y = b$. (If you substitute anything except a for x or anything except b for y the equations become false and so the logical product is false.) You go on for 2-classes:

$$\overset{a\ b}{\ } \qquad \overset{c\ d}{\ }$$
$$x = a \,.\, y = c \,.\, \mathbf{v} \,.\, x = b \,.\, y = d$$

And so you go on to classes of any number. And so we get to the surprising fact that all classes of equal number are already correlated one-one.[1]

Suppose that we substitute "eats" or "loves" for "="; and we take it for granted that everyone only eats himself or only loves himself.

Then: 'x eats a . y eats b' establishes a one-one correlation between a and b.

There is something queer here. But there is a strong temptation to make up such a relation.

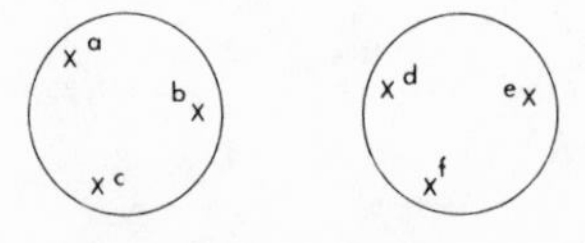

Now there is the same number of crosses in each of these rings. And somehow before I have correlated them they are in a way by their mere existence correlated. This is clearer perhaps where there is only one:

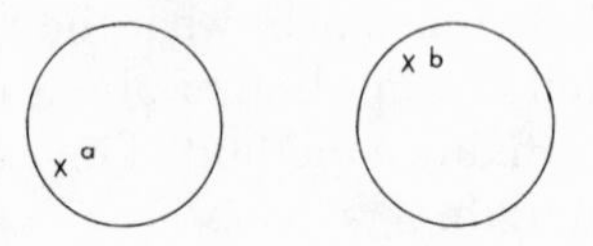

One wants to say, "You need not correlate a and b; they are already correlated."

1. Cf. *Philosophical Grammar*, pp. 355–358.

It is the relation existing between two things if the one is *a* and the other is *b*, the relation which they stand in by one being Turing and the other being Wittgenstein.—It is surprising that this should be called a relation; one is inclined to say, "And now let's have a relation."

What is the relation between two things if one eats Turing and the other eats Wittgenstein? Say two lions. Or two dogs, one of whom bit Turing and the other of whom bit Wittgenstein. There might be several couples of dogs for which this was true. So that T and W—or *a* and *b*—would be as it were the two "test bodies" for *x* and *y*, and *x* and *y* would have their relation by the one biting *a* and the other biting *b*. We could go about and ask, "Which two dogs have this relation?"

But what if I said, "What two people have the relation of the one eating himself and the other eating himself?"—if all people eat themselves? What would be the test here? You couldn't use this—the one eats himself and the other eats himself—to find out if two people had the relation.

You could use this—the one bites Wittgenstein and the other Turing—to correlate two classes. How would you use this relation—the one loves himself and the other loves himself—to find out whether they had the same number? [2]

Under what circumstances would you say that that relation holds (let's suppose we use it as a criterion for two sets of people having the same number)?

"Well, you name them all, and then write down this:

x loves a . y loves b."

True. That is what you do. And by giving them names and writing this down, you do correlate them. But this doesn't give you a relation by which you can establish that two classes have the same number.

It comes to saying this: "The class which contains *a* alone and the class which contains *b* alone have the same number. The class which contains *a* and *b* alone and the class which contains *c* and

2. (From "You couldn't".) Wittgenstein probably specified *which* relation he meant by pointing to the blackboard and saying "this". The specifications given in the text involve guesses based on all three accounts.

d alone have the same number." (This is connected with the fact that in logic the examples are always of classes which contain very small numbers.)

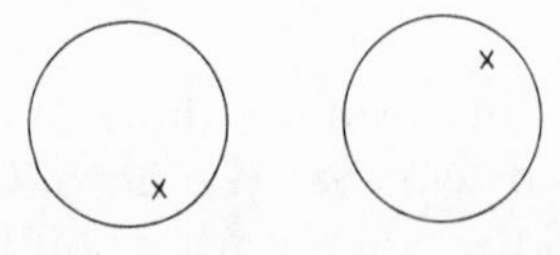

We actually say, "Well *this* is one and *this* is one." It is very important for the treatment in *Principia Mathematica* that there are classes whose numerical equality we can take in at a glance. We can write numbers | || |||, up to quite large numbers, especially with patterns. This ◯ ◯ is the model. Only the model just doesn't go on.

This shows the numerical equality of classes only if the numerical equality is an internal property and not an external property.

You may say: That the number of sides of these two triangles

is equal is an internal property of the triangles. For we make the number a criterion for the triangle, and the triangle a criterion for the number.

The business of naming things and correlating names depends on knowing when you have to repeat the same name; you must know under what circumstances to say you haven't given one thing two names. When my shadow coincides with Smythies's shadow, how many shadows are there and how many names must we give if we are naming shadows?

Take Russell's relation which holds between two things if the one is Lewy and the other is Wittgenstein.—When is one thing Lewy and one thing Wittgenstein?

Russell says that a thing is Lewy if it has all its properties in common with Lewy. Now Euclidean geometry is based upon the fact that people do measure—do discover the lengths of things; but what is it like to discover that something has all its properties in common with Lewy? How does it help us to say that a thing has all its properties in common with Lewy? To talk of this is hell and nothing else.

Leaving Russell's definition of equality—what does this sentence mean: that two things stand in the relation that one is Lewy and the other Wittgenstein? What's it mean except "There are two things, of which one is Lewy and the other Wittgenstein"?

I wouldn't even know what this meant: "*of* which one is Lewy". Is to be Lewy a property of a thing?

What test are we to apply in order to see whether in one box there are two apples and in another box also two apples? or only one apple in each box. You might say, "Let's see whether each apple is identical with itself." This wouldn't get you anywhere.

"Let's see whether this relation which Russell talks of really holds between the apples here." First of all, what is the relation of which Russell says it holds? It is either $x = a \,.\, y = b$; or $x = a \,.\, y = b \,.\mathbf{v}.\, x = c \,.\, y = d$; or $x = a \,.\, y = b \,.\, \mathbf{v} \,.\, ------- \,.\, \mathbf{v} \,.\, ------$; and so on. Now how am I to decide *which* of these to try?

Suppose that there is one apple in one box and two apples in the other. We try the second relation. "We find that it does not fit." But how do we find that? Well, suppose we give them all names and correlate the names. The first relation would be all right if we gave both the apples in one box the name "*a*".—We don't know how to apply a name to an apple, whether to apply two names "*a*" and "*b*" to an apple. You might say, "Oh, we mustn't do that." But how are we to find out whether we are doing that, except by counting?

Suppose we write one letter on each apple in one box, and then do the same to the other. The apples in one box go from *a* to *l* and in the other from *m* to *u*.—You might say that by this method you can *calculate* whether two classes have the same number, but you can't *measure* whether they have the same number.

Suppose we had to find out whether the letters from *a* to *h* have the same number as the letters from *o* to *v*. We could write down the formula given and see if there is a remainder or not; this is calculating whether they have the same number. Now why is this possible here and not with apples? Because here we have an internal property. We say, "The number of letters from *a* to *h* is so-and-so"—and this is timeless. If we said that there are different numbers of letters between the two at different times, then we could do nothing with the formula or method. It is a *calculation* if anything.

We could use it to determine, for example, "what class of cardinal numbers beginning with 15 has the same number as the class from 1 to 5?" We then begin with "$x = 1 \,.\, y = 15$", and so on.—You might therefore say that one of the Russellian relations holds between the cardinal numbers from 1 to 5 and the cardinal numbers from 21 to 25. This would then be a mathematical proposition. And "The relation holds between these classes" means that in order to correlate the classes we use this technique; and this technique is a technique of calculation. And even as it is, it is not applicable to large numbers; and a new technique is a new technique. Compare calculating whether something is a tautology or not.

Suppose we say

$$\sim\sim p \supset p$$
$$\sim\sim\sim\sim p \supset p$$
$$\sim^{(10^{10})} p \supset p$$

Whether this last is so or not is no longer found out in the way this is: $\sim\sim\sim\sim p \supset p$. We don't know what would be meant by "finding it out in that way". (You can have an image of a lot of "$\sim$"s merging into a haze; that's fine, but what can you do with it?)

In the definition of number, Russell and Frege made one great step—a colossally difficult step that had to be taken. Frege defined a number as a property of a property. It is not a property of a heap of apples. But it is a property of 'the property of being an apple lying on this chair'. This made one thing very clear: the relation between number and property.

Suppose we come into a room whose floor is littered with books. We try to arrange the books, and start by putting two apart: "Here are two volumes which certainly *don't* belong together." Does that mean that the books will remain where we put them? Not at all. But does that mean that this step is of no importance? Of course not.—[When we talk about what Russell and Frege were doing] I will constantly say, "and this again is muddle." But the value of it has to be borne in mind when I say this.

This business of "the property of being a man on this sofa" is a terrible muddle.—"This is a horse." But "Here is an x which is a horse"—no. The idea of all these things being predicates—"man", "circle", etc.—is a mass of confusions, and is of course embedded in Russell's symbolism:

$$(\exists x) \ldots$$

This comes from the English "There is a man such that . . ." But nothing in English provides for "There is an x which is a man." What would you say it is that is a man?

You can say, "This is a chair", meaning that it won't collapse if you sit on it, it is not made of paper, or some such thing. But normally "a chair" is not used as a predicate. We can say, "The only thing in the room is a chair." But not "There is an x which has the property of being a chair in the room." What that comes to is just "There is a chair in the room."

It is all right to say, "There is a pair of trousers which is grey." But not "There is a thing having the property of being a pair of trousers." Although I might say, "What you see is a pair of trousers."

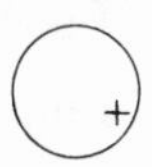

"The only thing in the circle is a cross." But if you asked, "What is the thing which is a cross?"—it might be all sorts of things: pieces of chalk, perhaps. Or we might answer, "What I have just drawn." As it is, we don't know what is referred to by "the x which is a cross".

And "*All* x's in this circle are crosses" is worse.

"Cross" looks like a predicate in certain contexts. But "(x): x

is in the circle. $\supset .x$ is a cross"—what does this mean? What is the *x*? There *are* sentences looking like this:

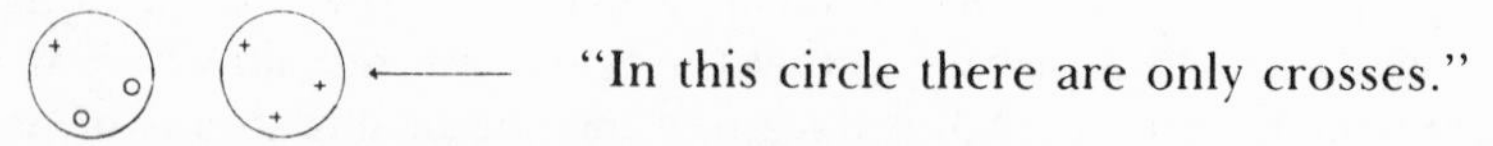

Only crosses? But there are also bits of cross, and white, and what not. Yet in this case it is clear what is meant: that all the figures I have drawn in it are crosses. And that makes sense.

What we normally mean by number is not at all always a property of a *property*. Because we would not know what has that property. Yet Frege's definition has made an enormous amount clear.

Frege went on to talk about the number of cardinal numbers. He called it "the number endless". This was the property of being one-one correlated to all the cardinals.

Suppose you had correlated cardinal numbers, and someone said, "Now correlate *all* the cardinals to all the squares." Would you know what to do? Has it already been decided what we must call a one-one correlation of the cardinal numbers to another class? Or is it a matter of saying, "This technique we might call correlating the cardinals to the even numbers"?

Turing: The order points in a certain direction, but leaves you a certain margin.

Wittgenstein: Yes, but is it a mathematical margin or a psychological and practical margin? That is, would one say, "Oh no, no one would call this one-one correlation"?

Turing: The latter.

Wittgenstein: Yes.—It is not a mathematical margin.

It seems as though when Frege introduced "the number endless" he had also told us how to count with it—what things have it. The queer thing is that as far as Frege is concerned, we have a number that is introduced only in mathematics. The other numbers occurred in mathematics but also outside mathematics. Or should we say, "No, but Frege only talked about the number 'endless' in mathematics; it was merely that he was not interested in its extra-mathematical use"?

If one had asked Frege, "What classes *have* the number 'endless'?", he would have replied, "The cardinals, the fractions, algebraic numbers, etc."—This doesn't show us at all in what English sentences the word 'endless' will be used. It is in fact used in ordinary life, but it plays a role quite different from what you expect.

Suppose that in Paris they not only keep the standard metre but also an exceedingly complicated structure used for comparing the metre rod with the foot rod [. . .] [3]

Turing: Does this complicated structure correspond to the method of proof that the number of, say, fractions equals the number of cardinals?

Wittgenstein: Yes, it does.

The point is that Frege hasn't told us what has the number 'endless'. You were led to think that probably if it were used at all it would be used for an immense collection of things. "The number of cardinal numbers" looks like "the number of a row of trees"—whereas we use it in sentences like "Jackie already knows endless (or $\aleph_0$) multiplications."

Professor Hardy says that the fact that there is no mathematician who has completed $\aleph_0$ syllogisms is of no more logical importance than the fact that there is no mathematician who has never drunk a glass of water.[4] This is a ghastly misunderstanding. The idea which you get is that the transfinite cardinals *are* not yet applied, and that if they were, they would have to be applied to something we can't reach.

Whereas: they *are* applied. They have a perfectly ordinary application, but not the application which Hilbert said.[5] For instance, I tell you, "Write down the first few terms of an $\aleph_0$"; and then you will perhaps write down "1, 2, 3, 4, . . ." or "1, 4, 9, 16, . . ."

3. There is only one version of this paragraph; it appears to be incomplete.

4. "Mathematical Proof", p. 5.

5. Hardy's remark, quoted above, was a reply to Hilbert's statement that no mathematician has completed an infinity of deductions. He was attacking the view he ascribes to Hilbert and Weyl that (as he puts it) "it is only the so-called 'finite' theorems of mathematics which possess a real significance." Cf. Hilbert, "On the Infinite", p. 151: "Our principal result is that the infinite is nowhere to be found in reality."

Or: "Go on building different streets as far as you can. But one thing: number the houses in each one with a different $\aleph_0$." This is all right.

But *not* "There are $\aleph_0$ trees in this row."

Nor "Lewy will never write down $\aleph_0$ syllogisms." What would this be like? We have no criterion. Even if I'd said "ten billion syllogisms", you could ask, "How do I find out? What's the way of counting in this case? Or do you mean roughly this sort of thing . . . ?" But "Write down an infinity of syllogisms"—the point is not that you can't do it, but that it *means* nothing.

I can say, "Ask any sum you like: I give you an $\aleph_0$ choice." But you can't then say, "Give me $\aleph_0$ shillings"; this would not mean anything.

Lewy: We do sometimes say that so-and-so has an endless amount of money.

Wittgenstein: Yes; in fact we might say that a certain bookkeeper has done an endless number of calculations. (But compare this with saying that Johnnie can do $\aleph_0$ multiplications.) But Professor Hardy and Hilbert both think that $\aleph_0$ is to be applied not to any actual bookkeeper but to a possible bookkeeper.

This word "$\aleph_0$" has nothing mysterious about it. But it plays a role quite different from what Hilbert and Professor Hardy think.

Hilbert translates from the mathematical role of $\aleph_0$ to the non-mathematical role, as he would from the mathematical role of 4 to the non-mathematical role of 4. But the non-mathematical role of $\aleph_0$ is quite different from the non-mathematical role of 4.

You begin by calling something a one-one correlation, say between even cardinals and cardinal numbers. This is like saying, "I'm writing down the series of cardinals" and writing "1, 2, 3, 4, . . ." This is quite all right. But you've only written down four numerals and some dots. The dots introduce a certain picture: of numbers *trailing off* into the distance too far for one to see. And a great deal is achieved if we use a different sign. Suppose that instead of dots we write Δ, then "1, 2, 3, 4, Δ" is less misleading.

Similarly with ". . . and so on". There are two ways of using the expression "and so on". If I say, "The alphabet is A, B, C, D, and so on", then "and so on" is an abbreviation. But if I say, "The cardinals are 1, 2, 3, 4, and so on", then it is not.—Hardy speaks as though it were always an abbreviation. As if a superman would write a huge series on a huge board—which is all right, but it has nothing to do with the series of cardinals.

What we have to see is not what role it plays in mathematics; because this suggests a wrong picture.

If these numerals, "$\aleph_0$", "$\aleph_1$", . . . , were introduced into an English grammar, you would see that "$\aleph_0$" is an entirely different part of speech from what you would expect it to be [from its role] in mathematics. And "$\aleph_1$" is a different part of speech again. And again with phrases like "greater than" as applied to these.

"Does Jackie know more multiplications than he knows cardinal numbers, or as many, or less?" You would explain that if he can go on indefinitely in each case, then we say the same.—But if I say, "I know the same number of calculations as Turing", this is already queer. Would you say of a man who knows one hundred kinds of calculation and a man who knows only one kind that one knows as many as the other? This would go against the grain—it would be a use of "as many as" which no one would ever use.

If you want to know what part of speech it is, go back to the wallpaper example. The master *doesn't* say to his apprentices, "Write down $\aleph_0$ curlicues", but rather: "You and you write down two different $\aleph_0$'s." That's why I gave the example of the wallpaper: it is a good way of finding out how "$\aleph_0$" is applied—what part of speech it is.

XVIII

Last term I said that Russell could not prove that $10 \times 100 = 1000$. What I ought to talk about now is the role that logic plays in mathematics, or the relation supposed to hold between logic and mathematics.

We came across the idea that although in Russell's symbolism, you could not prove the propositions of arithmetic, it is just a matter of giving the right definitions, and then Russell should be able to prove any proposition of arithmetic.—This is due to an idea one has about logic, that logic should be, as one might say, in no way arbitrary. In mathematics you might say, "Such-and-such a proposition is true, *given* that such-and-such axioms hold." But in logic we ought not to say such things. The whole essence of Russell's view is that there is only one logic. There must not be a Russellian and a non-Russellian logic, in the way in which there is a Euclidean and a non-Euclidean geometry.

Or if someone objected that "There is a Russellian logic and a non-Russellian logic"—then we might say, "All right, but then we won't call either of them logic at all. We must go a step further back in order to find something solid which underlies both."

One might say that although Russell's axioms are false, yet his way of deducing is the right way, and that is the solid foundation we are looking for; that is logic. It was this which made Russell say in *Principles of Mathematics* that all propositions of logic are of the form "if p then q". The question whether p is true, we could not prove. But "if p then q" we could prove.[1]

When it is held that logic is *true,* it is always held at the same time that it is not an experiential science: the propositions of logic are not in agreement or disagreement with particular experiences. But although everyone agrees that the propositions of logic are not verified in a laboratory, or by the five senses, people say that they are recognized by the intellect to be true. This is the idea that the intellect is some sort of sense, in the same way that seeing or hearing is a sense; it is the idea that by means of our intellect we look into a certain realm, and there see the propositions of logic to be true. (Frege talked of a realm of reality which does not act on the senses.) [2] This makes logic into the physics of the intellectual realm.

1. § 5.
2. *Grundgesetze,* I, xviii.

In philosophical discussions, you continually get someone saying, "I see this directly by inspection." No one knows what to say in reply. But if you have a nose at all, you will smell that there is something queer about saying you recognize truth by inspection.

What is the answer if someone says, "I see immediately that (say) $2 + 2 = 4$? Or that he is immediately aware of the truth of the law of contradiction? What should we say? Are we to take it lying down? It seems unanswerable; for how can you contradict such a person without calling him a liar? It is as if you asked him what colour he sees and he said "I see yellow."—What can you say?

Turing: One might ask him whether he can check it in any way.

Wittgenstein: Yes. And what if he says "No, I can't"?

Turing: One might then say that it does not matter much whether it is true or false.

Wittgenstein: Yes. We might ask, "Of what interest can it be that you say you see this?"

Suppose one shows a man a blue book and he says that he sees it yellow. Is it clear what consequences we have to draw?

Lewy: It is not clear, since we do not know, for example, whether he is claiming that the book is really yellow, and so on.

Wittgenstein: Yes. We might ask whether, if Lewy says that he sees blue, would that contradict the other's statement? And there is a possibility you can't rule out that he may be using the word "yellow" in a different way.

Am I trying (perhaps in a psychological laboratory) to find out how he uses the word "yellow", or am I trying to find out what colour he sees? Under special circumstances—say, I am trying to find out something about rays of light—his answer has a particular value. But otherwise—in other circumstances—it may have no value at all.

Suppose he says "This is immediately certain"—it is imagined that if he just utters these noises, then we know where we are. But we don't know at all. We don't know what consequences to draw. We don't even know if it is a joke or what it is. Only under very special circumstances do we know where we are.

Similarly if a man says that he sees as self-evident the law of contradiction. It might be the result of a psychological experiment, or alcohol. This doesn't as yet help us at all—unless we know what exact use is going to be made of this proposition. (If a medical student told his tutor he knew the whole of anatomy by intuition, he'd get the answer, "Well, you'll have to pass the examination like everybody else.")

Saying of logic that it is self-evident, meaning it makes a particular impression, doesn't help us at all. For one might reply, "If it is self-evident to you, perhaps it's not self-evident to someone else"—thus suggesting that his statement is a psychological one. Or we might ask, "What's interesting about your statement?"—thus suggesting the same thing.

So if we want to see in what sense the propositions of logic are *true,* what should we look for?—Ask what sort of application they have, how they are used.

How is one to know that the law of contradiction is true? We might ask: if we assume that the law of contradiction is false, what would go wrong?

Now what would it be like to assume that the law of contradiction is false?

Lewy: I might say, "Get out of this room and don't get out of this room", and expect you to act accordingly.

Wittgenstein: Yes. But there is something fishy here. What if I just stay leaning against the mantelpiece?

Can you understand such an order?—Suppose I gave you an order with a word you didn't understand—"Bring me an abracadabra." There also you would not understand; you might ask "What do you mean?", etc. But this doesn't look like our case of "Get out and don't get out". Or is it like it? Should we say simply that Lewy is talking nonsense and only making noises? Or is there something more to it than that?

Mme. Lutman-Kokoszynska: There is something more to it, because it is impossible to obey the command.

Wittgenstein: If we said this, we must distinguish it from the case where we are told to lift a very heavy weight. To say that it is impossible suggests that I am trying my hardest, but that I am unable to do it.

Von Wright: One might ask for rules according to which one was to obey it.

Wittgenstein: Well, suppose we teach a man to obey orders like "Bring me so-and-so". We teach him a simple language consisting of orders, the verb being "bring me", and then there are substantives: "apple", "book", etc. We teach him the names of these things by saying to him, "This is a book", etc.; and then later if we say, "Bring me a book", he brings a book. Also there is the word "not" and the word "and". Put whatever you like on the sides of "and": "Bring me so-and-so and bring me such-and-such", "Bring me so-and-so and don't bring me that other", and he always knows what to do. Except in the case of "Bring me a book and don't bring me a book".

We have taught him a technique. He hasn't been provided with any rules in this case. He wants a new rule of behaviour.—But now it seems as if he ought to know what to do in this case also.

We might want to say: What's wrong with the order? Or: Why doesn't the contradiction work?—Does it make sense to ask this question? Can one just say, "Well, it doesn't work, and that's all"?

In giving a contradictory order, I may have wanted to produce a certain effect—to make you gape, say, or to paralyze you. One might say, "Well, if this effect is what is wanted, then it does work."—People have thought it doesn't work because it produces this effect.

What sort of reasons could one give for why a contradiction doesn't work? Or am I making a mistake in asking this question?

Turing: In more complex cases one may ask this question when one wants the complexity unravelled.

Wittgenstein: Yes, that is the case if one wants to have it reduced to something else; for example, you show that it does not work *because* there is a contradiction. But the queer thing is that you say, "Surely a contradiction *can't* work."

In a sense, it is untrue to say it doesn't work; for if we gave rules for behaviour in the case of a contradictory order, then everything would seem to be all right. For example, "Leave the room and don't leave the room" is to mean "Leave the room hesitatingly". Can one then say the contradiction works perfectly? Have we given the contradiction a sense, or not?

Lewy: One might say that an entirely new meaning has been given to the contradiction.

Wittgenstein: Yes, one might say that.—And notice that contradictions are actually often used in this way. For instance, we say, "Well it is fine and it's not fine", meaning that the weather is mediocre. And one might even introduce this use into mathematics.

Suppose that we give this meaning to contradictions. Then the order "Go out and do not go out" might work in many cases—might produce the right response. Lewy says that we have then given the contradiction an entirely new meaning.—We might say first that for some purposes this would be most inconvenient. And also: What is an entirely *new* meaning? Is it clear what is an old and what is a new meaning?

Think of "going on in the same way". Suppose that I am taught to move one, two, or three paces forward when one, two, or three fingers are held up; when four fingers are held up, I am taught to climb onto the chair. Is my climbing onto the chair a new thing or not?—[Suppose] someone then said that climbing the chair was not consistent with the first three things I was taught. But isn't it consistent? It is not consistent with the formula which prescribes that one should go one pace forward for every finger which is held up; but couldn't we make a formula in which 4 has an isolated position? The point is that there is no sharp line between a regular use and an irregular or capricious use. It wouldn't even be a capricious use if one day you did it in one way, another another.

If you say it's a new meaning, this isn't clear. What is clear is that if I taught you this technique and then gave you the contradiction, you wouldn't know what to do.—I might use the contradictory orders as a sort of decoration, an extra ornament of the language, just in order to fill in time. I might not want you to do anything.

On the other hand we can imagine people who had learned the technique, and who, when they were given the order "Bring me a book and do not bring me a book", would do something in such a way that we'd say, "They take it for granted they are following the order."

Let us go back a bit. At first sight we want to ask why a contradiction does not work. But I might say that there isn't really any explanation at all. Or rather, no—this is incorrect, too. We can give explanations. But we have to ask what these explanations do for us.

I once wrote the law of contradiction and other propositions of logic in the form of a certain symbolism; and I regarded this as a sort of explanation.

I tried to explain the self-evidence of logical propositions by writing down schemata like this:

p	q	
T	T	T
F	T	F
T	F	F
F	F	F

This was given as another way of writing the proposition 'p and q'; and assuming a certain [order] of permutations, we can write it as TFFF (p,q).

Incidentally, this kind of schema is not my invention; Frege used it.[3] The only part of it which is my invention—not that it matters in the least—was to use this as a symbol for the proposition, not as an explanation of it (like Frege).

If you write '$p. \sim p$' in this symbolism, you get a proposition which has only F's. Then: '$\sim (p. \sim p)$'—we get a proposition, the law of contradiction, which has only T's; that is, we show that the law of contradiction is true in all cases. We can then show that Russell's primitive propositions are chosen in this particular way—they are tautologies.[4]

You might say that this symbolism gives an explanation of why

3. See, e.g., *Begriffschrift,* § 7.

4. (From "If you".) The passage is based on B and R; but their accounts are very different.

a contradiction doesn't work, and of why a proposition of logic may be said to be true, but is not verifiable by experience.

One could make this analogy. Suppose we had a mechanism consisting of four cogwheels:

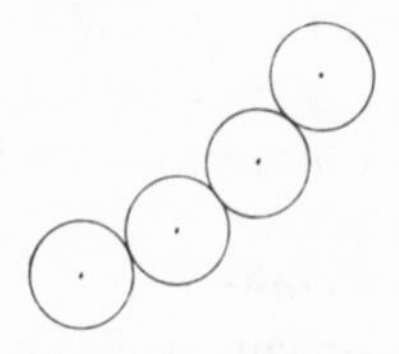

We would then be accustomed to the fact that if we turned one cogwheel here, its movement determines the movement of the last cogwheel of the chain. I might then show you that there are mechanisms with cogwheels, such as the differential gear of a car, in which you can turn one cogwheel and at the same time you can do with another cogwheel just what you please. Here there is a pseudo-connexion; the connexions are cancelled out and you can do what you like.

Then there is another mechanism with cogwheels, very simple:

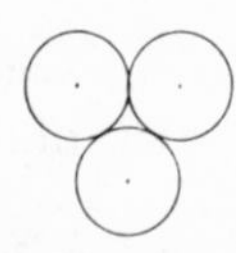

This one cannot move at all. You might say this is like a contradiction and the differential gear is like a tautology. For the triangular system of cogs and the differential both look like mechanisms; in both cases you have connexions—but in the former case you can do nothing with it, and in the latter, you can do anything with the other wheel you like. Similarly you might say that if you give a man a contradictory order, he has no room to move at all; and if you give him a tautological order ("Leave the room or don't leave the room") he can do anything he pleases.

Now in what sense is this an explanation? "A contradiction jams and a tautology does nothing"—have I now explained why a contradiction doesn't work? Have I explained by means of my symbolism why logic is true?

Lewy: You have substituted one symbolism for another.

Wittgenstein: Yes.

I said that a contradiction jams, and this sounds very good. But what the hell does it mean, saying it jams? All that happened was that I said, "Bring me a book and don't bring me a book", and Lewy just stood there; he didn't know what to do. But is this the jamming? If so, it is a *psychological* jamming. But when I said "it jams", we thought it meant a logical jamming, not a psychological jamming. For we feel that it is not Lewy's fault that he did not know what to do. If he had done something, we would have said, "This isn't the original meaning", whatever he did. We would not have taken anything to be the correct fulfilment of the order.

When we say it jams, we don't mean simply the fact that people don't react correctly. But we expect a man who knows the language to say, "This makes no sense." Or we could put it: If we have a certain number of orders of a certain kind, and then such orders connected by "and" and "not", then we would recognize certain actions to be the fulfilling of certain orders, and we would not recognize *any* action to be the fulfilling of the contradictory order.

There are all sorts of reasons for this. For instance, we may say it would be extremely *inconvenient* to give the contradictory order a meaning.

What I am driving at is that we can't say, "So-and-so is the logical reason why the contradiction doesn't work." Rather: *that* we exclude the contradiction and don't normally give it a meaning, is characteristic of our whole use of language, and of a tendency not to regard, say, a hesitating action, or doubtful behaviour, as standing in the same series of actions as those which fulfil orders of the form "Do this and don't do that"—that is, of the form '$p. \sim q$'.

This is connected with the problem whether we ought to say that a double negation is equivalent to an affirmation or a negation. In some languages, a double negation is a negation. —Suppose I say, "Russell chooses $\sim\sim p \equiv p$, but I choose $\sim\sim p \equiv \sim p$". We might ask: how is this possible? If people sometimes use a double negation as a negation, can one say that they are wrong?

Von Wright: Could we say: in one sense of negation $\sim\sim p \equiv p$, and in another sense of negation $\sim\sim p \equiv \sim p$?

Wittgenstein: Well, yes, but there is something queer about this. For it may mean one of two things. It may mean that there is a sense (a) in which $\sim\sim p \equiv p$ and another sense (b) in which $\sim\sim p \equiv \sim p$. Or it may mean that $\sim\sim p \equiv p$ or $\sim\sim p \equiv \sim p$ makes the "$\sim$" be used in different senses in the two cases; they each define a sense of negation. Is saying that Russell uses sense (a) simply the same as saying that Russell uses "$\sim$" so that $\sim\sim p \equiv p$? Or does using "$\sim$" in sense (a) *produce* $\sim\sim p \equiv p$? Telling me that in one sense $\sim\sim p \equiv p$ and that in another sense $\sim\sim p \equiv \sim p$ is to tell me nothing, unless you say what the senses are.

Couldn't we explain $\sim\sim p \equiv p$? We might suggest, "If 'not' is regarded as a reversal—as a turning round 180°—then: The ruler points to him; negate, and it points away from him; negate again in the same way, and it points to him as before." One might even have a notation in which one wrote *p* upside down to signify not-*p*.—On the other hand one might regard double negation as first turning a thing, and then taking it back to its original position, and then turning it again, for the sake of emphasis. And then $\sim\sim p \equiv \sim p$. In this case one has the diagram  instead of the diagram  to signify double negation.

Now does all this constitute an explanation or not? Is saying that negation is a reversal an explanation? Isn't it similar to the T-F notation?

You might think that you can explain the two uses of double negation by means of brackets, writing one of them as '$(\sim\sim) p \equiv \sim p$' and the other as '$\sim(\sim p) \equiv p$'. A bracket seems to explain a lot; but why should it? Brackets are simply dashes; they are symbols as much as anything else.

We want to say that the brackets in '$\sim(\sim p)$' mean "Do the same thing with $\sim p$ as you've done before with *p*." But "do the same thing"? Who says what "the same thing" is? Suppose that one turns a chair round and is then told to do the same again. What is "the same" here? Is one to turn it back into its original position or is one to put it in its original position and turn it again? Must this be clear? Or isn't it a question of: "How are we most likely to react?"

One might say that the brackets in '$\sim(\sim p)$' help the understanding. But that only means that people will normally react to them in such-and-such a way. Similarly the figure: ⟲ may help the understanding. But it is only a figure, and the important question is how we are going to use it.

When we say, "Given a certain sense of 'not' or of 'doubling the negation', it is clear what the result will be", this may mean two things. (a) It may mean only that you *call* getting such-and-such a result "using double negation in this sense", etc. (b) It may mean that if we associate a certain picture with double negation we are more likely to do *this;* if we associate a different picture, we are more likely to do *that.* In this case it is perfectly all right to talk about "one *sense* of double negation"—referring to the picture and the inclination that goes with it—and "another sense". The bracket is [such] a picture. But of course no picture compels us to get a certain result, since any picture can be used in all sorts of different ways.

Similarly, the T-F notation is a picture which we can hardly associate with any other kind of usage. But it *could* again be reinterpreted. And it does not show at all that if we have a contradiction with the symbol FFFF in this notation, then this could not be given sense.

I should like to show that one tends to have an altogether wrong idea of logic and the role it plays; and a wrong idea of the *truth* of logic. If I can show this, it will be easier to understand why logic doesn't give mathematics any particular firmness.

XIX

What would go wrong, if anything, if we didn't recognize the law of contradiction—or any other proposition in Russell's logic? We treated the question of double negation as parallel to that: If some people used double negation to mean affirmation, and others used double negation to mean negation, should we say then that they were using negation—or double negation—with "different meanings"? We discussed whether a particular meaning of negation made a certain usage correct, or whether that meaning consists in using negation in that way.

This is a difficulty which arises again and again in philosophy: we use "meaning" in different ways. On the one hand we take as the criterion for meaning, something which passes in our mind when we say it, or something to which we point to explain it. On the other hand, we take as the criterion the use we make of the word or sentence as time goes on.

First of all, to put the matter badly and in a way which must be corrected later, it is clear that we judge what a person means in these two ways. One can say that we judge what a person means by a word from the way he uses it. And the way he uses it is something which goes on in time. On the other hand, we also say that the meaning of a word is defined by the thing it stands for; it is something in our minds or at which we can point.

The connexion between these two criteria is that the picture in our minds is connected, in an overwhelming number of cases—for the overwhelming majority of human beings—with a particular use. For instance: you say to someone "This is red" (pointing); then you tell him "Fetch me a red book"—and he will behave in a particular way. This is an immensely important fact about us human beings. And it goes together with all sorts of other facts of equal importance, like the fact that in all the languages we know, the meanings of words don't change with the days of the week.

Another such fact is that pointing is used and understood in a particular way—that people react to it in a particular way.

If you have learned a technique of language, and I point to this coat and say to you, "The tailors now call this colour 'Boo' ", then you will buy me a coat of this colour, fetch one, etc. The point is that one only has to point to something and say, "This is so-and-so", and everyone who has been through a certain preliminary training will react in the same way. We could imagine this not to happen. If I just say, "This is called 'Boo' " you might not know what I mean; but in fact you would all of you automatically follow certain rules.

Ought we to say that you would follow the *right* rules?—that you would know *the* meaning of "boo"? No, clearly not. For which meaning? Are there not 10,000 meanings which "boo" might now have?—It sounds as if your learning how to use it were different from your knowing its meaning. *But the point is that we all make the SAME use of it.* To know its meaning is to use it *in the same way* as other people do. "In the right way" means nothing.

You might say, "Isn't there something else, too? Something besides the agreement? Isn't there a *more natural* and a *less natural* way of behaving? Or even a right and a wrong meaning?"—Suppose the word "colour" used as it is now in English. "Boo" is a new word. But then we are told, "This colour is called 'boo' ", and then everyone uses it for a shape. Could I then say, "That's not the straight way of using it"? I should certainly say they behaved unnaturally.

This hangs together with the question of how to continue the series of cardinal numbers. Is there a criterion for the continuation—for a right and a wrong way—except that we do in fact continue them in that way, apart from a few cranks who can be neglected?

We do indeed give a general rule for continuing the series; but this general rule might be reinterpreted by a second rule, and this second rule by a third rule, and so on.

One might say, "But are you saying, Wittgenstein, that all this is arbitrary?"—I don't know. Certainly as children we are punished if we don't do it in the right way.

Suppose someone said, "Surely the use I make of the rule for continuing the series depends on the interpretation I make of the rule or the meaning I give it." But is one's criterion for meaning a certain thing by the rule the using of the rule in a certain way, or is it a picture or another rule or something of the sort? In that case, it is still a symbol—which can be reinterpreted in any way whatsoever.

This has often been said before. And it has often been put in the form of an assertion that the truths of logic are determined by a consensus of opinions. Is this what I am saying? No. There is no *opinion* at all; it is not a question of *opinion.* They are determined by a consensus of *action:* a consensus of doing the same thing, reacting in the same way. There is a consensus but it is not a consensus of opinion. We all act the same way, walk the same way, count the same way.

In counting we do not express opinions at all. There is no opinion that 25 follows 24—nor intuition. We express opinions by means of counting.

People say, "If negation means one thing, then double negation equals affirmation; but if it means another thing, double negation equals negation." But I want to say its use is its meaning.

There are various criteria for negation.—Think of the ways in which a child is taught negation: it may be explained by a sort of ostensive definition. You take something away from him and say "No".

A child is trained in a certain technique of applying negation long before the question of double negation arises. If a child is taught the use of negation apart from all this, and then goes on to use double negation as equivalent to negation, would you say he is necessarily using negation now to mean something different? If you say, "It *must* have a different meaning now"—this says nothing, unless you mean that a different picture will be associated with it.

Let us go back to the law of contradiction. We saw last time that there is a great temptation to regard the truth of the law of contradiction as something which *follows* from the meaning of negation and of logical product and so on. Here the same point arises again.

I will now use an awful expression. I wanted to talk of a stationary meaning, such as a picture that one has in one's mind, and a dynamic meaning. I was going to say, "No dynamic meaning follows from a stationary meaning." But that is very badly put and had better be forgotten immediately.

Another way of putting it is to warn you: Don't think any *use* collides with a picture, except in a psychological way. Don't imagine a sort of logical collision. But that is also very badly expressed. For one then wants to ask where I got the idea of logical collision from. And one would be perfectly justified in asking.

One is tempted to say, "A contradiction not only doesn't work—it *can't* work." One wants to say, "Can't you see? I can't sit and not sit at the same time." One even uses the phrase "at the same time"—as when one says, "I can't talk and eat at the same time." The temptation is to think that if a man is told to sit and not to sit, he is asked to do something which he quite obviously can't do.

Hence we get the idea of the proposition as well as the sentence. The idea is that when I give you an order, there are the words—then something else, the sense of the words—then your action. And so with "Sit and don't sit", it is supposed that besides the words and what he does, there is also the *sense* of the contradiction—that something which he can't obey.

One is inclined to say that the contradiction leaves you no room for action, thinking that one has now *explained* why the contradiction doesn't work.

Suppose that we give the rule that "Do so-and-so and don't do it" always means "Do it". The negation doesn't add anything. So if I say "Sit down and don't sit down", he is to sit down. If I then say, "Here you are, the contradiction has a good sense", you are inclined to think I am cheating you. This is an immensely important point. Am I cheating you? Why does it seem so?

Turing: I should say that we were discussing the law of contradiction in connexion with language as ordinarily used, not in connexion with language modified in some arbitrary way which you like to propose.[1]

Malcolm: The feeling one has was that we were talking of '$p . \sim p$' as it is now used—to express a contradiction; and you have merely suggested a use in which it would no longer express a contradiction.

Wittgenstein: Yes; you speak of the sentence as expressing a

1. There is a remark in S, "The only modification I suggested was a modification in *this* expression", which *may* have been a reply, or a part of a reply, to Turing.

contradiction—as if the contradiction were something other than the sentence and expressed by it.—But doesn't the explanation of this feeling that I have cheated lie perhaps in the fact that I have made a wrong continuation?

Now what is it that I have continued wrongly?

Turing: Could one take as an analogy a person having blocks of wood having two squares on them, like dominoes. If I say to you "White-green", you then have to paint one of the squares on the domino which I give you white and the other green. If the point of this procedure is to be able to distinguish the two squares, you will probably hesitate when I say "White-white". —Your suggestion comes to saying that when I say "White-white" you are to paint one of the squares white and the other grey.

Wittgenstein: Yes, exactly. And where does the cheating come in? What is the wrong continuation I have suggested? Why is this continuation in your analogy a *wrong* continuation? Might it not be the ordinary jargon among painters?

The point is: Is it or is it not a case of one continuation being *natural* for us? Or ought one to say that there is something more to it than that? Ought one to give a reason why one continuation is natural for us? Ought one to say this, for example: "If we learn to use orders of the form '*p*', '*q*', '*p* and *q*', '*p* and not-*q*' etc, then so long as we give the phrase '*p* and not-*p*' the sense which is determined by the previous rules of training, it is clear that this cannot be a sensible order and cannot be obeyed. If the rules for obeying these orders—for logical product and negation—are laid down, then if we stick to these rules and don't in some arbitrary way deviate from them, then of course '*p* and not-*p*' can't make sense and we can't obey it." Isn't that the sort of thing you would consider *not* cheating?

Turing: I should say that it is another kind of cheating. I should say that if one teaches people to carry out orders of the form '*p* and not-*q*' then the most natural thing to do when ordered '*p* and not-*p*' is to be dissatisfied with anything which is done.

Wittgenstein: I entirely agree. But there is just one point: does "natural" mean "mathematically natural"?

Turing: No.

Wittgenstein: Exactly. "Natural" there is not a mathematical term. It is not mathematically determined what is the natural thing to do.

We most naturally compare a contradiction to something which jams. I would say that anything which we give and conceive to be an explanation of *why* a contradiction does not work is always just another way of saying that we do not want it to work.

If you have a tube and a cock which shuts or opens it, your experience may have led you to think that always when the handle is parallel to the tube, the tube is open, and when it is at right angles to it, the tube is closed. But at home I have a cock which works the other way about. And in order to get used to it, I had to think of the handle as lying along the tube and blocking it, so that the tube was closed when the handle was parallel to it. I had to invent a new imagery.

Similarly, one needs to change one's imagery in the case of contradictions. One can change one's imagery in such a way that 'p and not-p' sounds entirely natural, as when we say, "The negative doesn't add anything". This is most important. We shall constantly get into positions where it is necessary to have a new imagery which will make an absurd thing sound entirely natural.

I want to talk about the sense in which we should say that the law of contradiction:

$$\sim (p . \sim p)$$

is a true proposition. Should we say that if '$\sim (p . \sim p)$' is a true proposition, it is true in a different sense of the word from the sense in which it is a true proposition that the earth goes round the sun?

In logic one deals with tautologies—propositions like '$\sim (p . \sim p)$'. But one might just as well deal with contradictions instead. So that *Principia Mathematica* would not be a collection of tautologies but a collection of contradictions. Should one then say that the contradictions were true? Or would one then say that "true" is being used in a different sense?

Turing: One would certainly say that it was being used in a different way.

Wittgenstein: It is used in a different way because you now say it of things of which you would not say it before.

One could put the point this way. One often hears statements about "true" and "false"—for example, that there are true mathematical statements which can't be proved in *Principia Mathematica,* etc. In such cases the thing is to avoid the words "true" and "false" altogether, and to get clear that to say that p is true is simply to assert p; and to say that p is false is simply to deny p or to assert $\sim p$. It is not a question of whether p is "true in a different sense". It is a question of whether we assert p.

If a man says "It is fine" and I say "It is not fine", I am correcting him and asserting the opposite; and we can then argue about whether it is fine or not, and we may be able to settle the question. But if I am trained in logic, I am trained to assert certain things and not to assert others. This is an entirely different case from being trained to assert that Smith looks sad. I am not trained to assert that he looks sad or that he doesn't look sad. But I am actually trained to assert mathematical propositions—that $3 \times 6 = 18$, and not 19—and logical propositions.

"Trained to assert"—under what conditions? Well, for instance, when I have to pass an exam.—And if, for example, we did logic by means of contradictions, we should be trained to assert contradictions in examinations.

It is important in this connexion that there is an inflexion of asserting. We make assertions with a peculiar inflexion of the voice; and there are gestures with this. This is one thing which is very characteristic of assertion. It is also important that assertions in our language have a peculiar jingle; we make them with sentences of a certain form. For instance, "'Twas brillig" is an assertion, although "brillig" is not a normal word.

Now suppose that we were trained to use contradictions instead of tautologies in logic. There are circumstances in which we should call it the same logic as our present logic. What are these circumstances? What would be our criterion for saying that this other logic is all absurd, or for saying that it is essentially the same as our present logic?

Malcolm: Wouldn't we say it was the same as our present logic if we used "$\sim$" in a different way?

Wittgenstein: Yes, But using "$\sim$" in a different way does not here refer to the way in which it is used in the proofs. [In the proofs it] might be just the same.—In ironical statements, a sentence is very often used to mean just the opposite of what it normally means. For instance, one says "He is very kind", meaning that he is not kind. And in these cases the criterion for what is meant is the occasion on which it is used.

One might make a deduction and say "He is very kind, therefore we will give him a birthday present" ironically, meaning "He is not kind, therefore we will not give him a birthday present." Thus we could have proofs in our supposed new logic just like the ones in *Principia Mathematica,* and the assertion sign would appear before contradictions.

By the way, this is the way in which a proposition can assert of itself that it is not provable. Besides putting the assertion sign before contradictions I could put it before propositions like '$p \supset q$'. In the one case '$\vdash p. \sim p$' would mean '$p. \sim p$ is refutable'; and in the other '$\vdash p \supset q$' would mean '$p \supset q$ is not provable'. Thus we see that *Principia* might not only be a collection of tautologies or a collection of contradictions; it might even be a collection of propositions which are neither contradictions nor tautologies.

In our ordinary logic we read '$\vdash \sim(p. \sim p)$' as 'It is the case that not (p and not-p)'. In the new logic of contradictions, we could read '$\vdash p. \sim p$' as 'It is the case that p and not-p' or 'It is true that p and not-p' or as just 'p and not-p'. Similarly in the third logic that we considered, you might read '$\vdash p \supset q$' as 'It is true that p implies q' or as 'It is the case that p implies q'. And you could say 'It is true that p implies q' or 'It is true that p and not-p' with just the same gestures and tone of voice as you now say 'It is true that p or not-p' for '$\vdash p \vee \sim p$'.—It is easy to see why in this new logic we are unwilling to read '$\vdash p \supset q$' as 'It is true that p implies q'. But '$\vdash p \supset q$' is the proposition which in that logic you read as 'p implies q'; and to add 'It is true that' or 'It is the case that' makes no difference. It doesn't commit you to any more than saying 'p implies q'.[2]

2. This paragraph is based on B, M, and S. B was apparently quite inaccurate, but was much fuller than the others. They have been used to correct the B account.

All that I wish to do by this is to show that there are all sorts of different ways in which we could do logic or mathematics. And the fact that we read it out and say every time 'It is true that' makes no difference. What matters is how we later use the things which we read out.

XX

One thing I tried to say last time can be said as follows. When one considers contradiction and feels the need of explaining why a contradiction won't work, one is inclined to speak of "the *mechanism* of contradiction". And in a similar way one might talk about the *mechanism* of negation or disjunction.

"What is it to negate a proposition?" one asks. "What happens when a proposition is negated? For surely something is *done* to it. It can't be just putting the word 'not' before it. There must be something else." And then it seems that putting "not" in front of the proposition is only a sign of some sort of activity that takes place—say, in one's mind—which is the negating; and one is inclined to ask what this is.

So we have the idea of a contradiction "jamming". And this is only another way of saying that the *meanings* of the signs jam. Professor Moore, in his paper to the Moral Science Club at the beginning of this term, wanted to say that in a contradiction the meanings jam in some sense.—I will try to show that the picture of a mechanism here is an extremely misleading one. It is in such pictures that most of the problems of philosophy arise.

The important point is to see that the meaning of a word can be represented in two different ways: (1) by an image or picture, or something which corresponds to the word, (2) by the use of the word—which also comes to the use of the picture.

Now what is it which is supposed to jam? The pictures or the use? Of the *use* you can't say it jams, because you have a right to fix the use as you like. But how could *pictures* jam? There is only one way in which they could, and that is a psychological way.

The phenomenon of jamming *consists* in the fact that we say it jams: that we say, "Oh, it's a contradiction and we cannot do anything with it", etc. The phenomenon is not, as it were, somewhere else and observed by us in some other sphere.

Another thing we are inclined to say is that if we allowed contradictions, we could not do certain things, or that we can't use language in a certain way. And thus Frege once said that if we denied certain logical laws—for example, if we did not admit the law of identity to be true—our thinking would become confused and we should have to give up making judgments.[1]—Here we have the same mistake coming in again.

Suppose I said that we have to recognize certain logical laws—certain rules about negation, for instance—because if we didn't, we could not use negation in a certain way.—But what is it that *defines* negation? What is it that characterizes negation as negation? If someone says "He is not here", we call that negation. But it is not the sound "not" which is negation; for the same sound might in Chinese mean "flowerpot".

What use of a word characterizes that word as being a negation? Isn't it the *use* that makes it a negation?

It is not a question of our first *having* negation, and then asking what logical laws must hold of it in order for us to be able to use it in a certain way. The point is that using it in a certain way is what we mean by negating with it.

We explain negation in a particular way—perhaps by taking a lump of sugar away from a child and saying "No". Then later we give other rules for negation—for example, the rule that two negations make an affirmation. Now somebody says, "Unless we recognized these rules we could not use negation as we do." What does it mean to say this? Is it correct to say it or not?

Turing: What is in one's mind if one says that sort of thing is something like this: One starts teaching the child negation by not allowing it to have sugar; but one does not yet formulate the

1. *Grundgesetze,* I, xvii.

logical rules. Then one applies the negation thus learnt to all sorts of propositions. And the idea is that the only natural way of applying it to all sorts of propositions is in such a way that these logical laws hold.

Wittgenstein: Yes; and let us take another example; the use of "all". "If all the chairs in this room were bought at Eaden Lilley's then this one was. $(x).\, fx$ entails fa." Suppose I ask, "Are you sure fa follows from $(x).\, fx$? Can we assume that it does not follow? What would go wrong if we did assume that?"

Wisdom: One reply which might be given is that it is impossible to make such an assumption.

Wittgenstein: Yes. But let us look into this, because such things as "Let us assume that $(x).\, fx$ does not entail fa" have been said. Now the reply you suggested did not mean that it is psychologically impossible to assume that; for if it did, one might say that although Wisdom cannot do it yet perhaps other people can.

In what way is it impossible to assume $(x).\, fx$ does not entail fa?

Wisdom: Isn't the assumption like saying "Couldn't we have a zebra without stripes?"

Wittgenstein: Yes. It would be said that the meaning of '$(x).\, fx$' had been changed.

What then would go wrong if someone assumes that $(x).\, fx$ does not entail fa? I would say that all I am assuming is a different use of "all", and there is *nothing* wrong in this.

If I stick to saying that the meaning is given by the use, then I cannot use an expression in a different way without changing the meaning. But it is then misleading to say, "The expression *must* have a different meaning if used differently." It is merely that it *has* a different meaning—the different use *is* the different meaning.

And if one says, "If one assumes fa does not follow from $(x).\, fx$, one must use $(x).\, fx$ in a different way"—we reply, in assuming this one *does* use it in a different way.—But if we make this assumption, nothing goes wrong.

One might say, "No, Wittgenstein, it does not work as you say. For if it were like that, there would be nothing revolting about assuming that *fa* does not follow from $(x). fx$." Then, in order to show what is revolting about it, you have to say something like: "It isn't true that it's just the use which defines the meaning. Rather '$(x). fx$' has a meaning—which this use you suggest does not fit."

Now where does this "does not fit" come in? For it is perfectly true that it does come in somewhere.

Let us see how we explain "all". I might explain "all the men in this room"—showing them all and making some suitable gesture; "all the bits of chalk"—pointing to each one. This is a picture which the word can call up. But then after explaining this, I might say, "All the men in this room are over 25, but *he* isn't." Suppose you then say, "Which are 'all the men'?"; and I point to each in turn, including him.—Now is there a contradiction in this? You see, one might explain the word in the same way we do, and have in one's mind the same picture,—and one might nevertheless use it in quite a different way. Only that would be highly unnatural to us.

Similarly, if I give a man a table of colour samples with the name "sea-green" under one of them, and then say "Bring me a sea-green book", it would be highly unnatural if, instead of looking round for a book the same colour as the sample, he were to look round for the complementary colour. But he might do so.

There is a very firm connexion between the way we learn a word and the way we use it. And in *this* sense we might say: This way of learning 'contradicts' this way of using; or: It 'contradicts' the meaning of "all" not to let *fa* follow from $(x). fx$. But it is here a matter of a peculiar picture being always connected with one use rather than with another use.

This is connected with the fact that there is, in all the languages we know, a word for "all" but not for "all but one". This is enormously important: this is the sort of fact which characterizes our logic. "All but one" seems to us a complex idea—"all", that's a simple idea. But we can imagine a tribe where "all but one" is the primitive idea. And this sort of thing would entirely change their outlook on logic.

We talked of the idea that if we did not recognize certain logical laws we could not do with negation what we wanted to do with it. But this is not correctly expressed. We might say: If we don't follow this rule, then the word isn't a negation—because we take this rule as essential for what we call negation. Or we might say: Yes, it is a negation, but a rather unnatural form of negation—nobody would ever use it. Like an arithmetic leaving out the number 5.—But one might find a people who left out 13 and had very complicated rules about that point. This wouldn't seem so unnatural, and there are facts which recommend it.

I am speaking against the idea of a "logical machinery". I want to say there is no such thing.

The idea of a logical machinery would suppose that there was something *behind* our symbols. Thus there are certain cogwheels behind the dial of a clock which produce the following movement: if I move the minute hand around once, the hour hand will move a twelfth part of the circle in the same direction. In the foreground we have nothing but the two hands which work in a particular way, which way is explained by the machinery in the background.

Similarly, one might think that there is a machinery behind the symbols—that behind '$(x).\ fx$' and 'fa' is a machinery which explains why one must follow from the other. A Chinaman who just sees the symbols wouldn't see this machinery. But we who see the machinery see that if there is $(x).\ fx$, there must be fa.

For us a machinery often stands as a symbol for a certain action.

If I wish to explain what the hour hand will do when I move the minute hand in certain ways, one way of explaining it is to take the back off the clock and show you the works. The machinery is actually used to explain the motions of the two hands.

There are other ways of explaining to you what the hour hand will do. For instance, I may turn the minute hand round, say, three times, and let you see the hour hand move a quarter of a complete circle. But you may not be able to predict from this what the hour hand will do if I turn the minute hand once more round.

Or you may be sceptical. ("It might do anything. For you can imagine a mechanism which will produce any movement you like during the next turn.") But if I show you the mechanism behind the dial, you will be able to predict the movement of the hour hand for any given movement of the minute hand; and you will not be sceptical. Showing you the mechanism is normally treated as a much more general explanation.

But isn't this queer—that a mechanism is treated as a general explanation? What do I show you when I show you a mechanism? I show you cogwheels and pins. Perhaps I don't even show you the mechanism moving.

The point is that just looking at the cogs would not by itself seem to give you more at all than moving the hands would—perhaps less. You might think the cogs would vanish away, or explode. But you don't. The fact is, we use the mechanism as a symbol for a certain kind of behaviour. We do this again and again. But you can't say we are making an *assumption* about what will happen to the mechanism. For instance, I may drop the clock so that the machinery is broken, or lightning may strike it—but one would not say that I had made any false assumptions.

It is simply one of our ways of explaining a kind of behaviour, to explain the mechanism. For instance, suppose I show you this figure

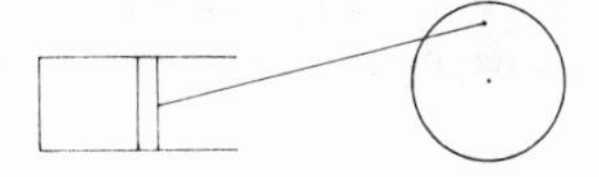

and ask you what will happen if I turn the wheel through 90° in an anticlockwise direction. Then you will all make such-and-such a construction, making the connecting rod equal in length to the connecting rod in the figure and you will produce this second figure

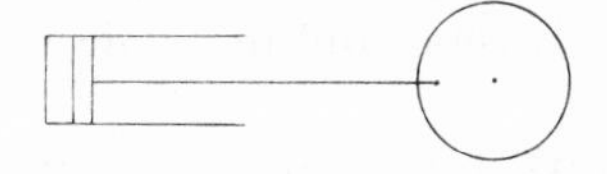

We use it as a rule of construction in cases like this that the connecting rod shall be of equal length always. And we can describe the movement of a thing by saying that it moves as it would if it were worked by such-and-such a mechanism.

If we talk of a logical machinery, we are using the idea of a machinery to explain a certain thing happening *in time.* When we think of a logical machinery explaining logical necessity, then we have a peculiar idea of the parts of the logical machinery—an idea which makes logical necessity much more necessary than other kinds of necessity. If we were comparing the logical machinery with the machinery of a watch, one might say that the logical machinery is made of parts which cannot be bent. They are made of infinitely hard material—and so one gets an infinitely hard necessity.

How can we justify this sort of idea?

One has in mind that branch of mathematics which is called kinematics (though the word "kinematics" may be used also in other senses). Kinematics is really a branch of geometry; in it one works out how pistons will move if one moves the crankshaft in such-and-such a way, and so on. One always assumes that the parts are perfectly rigid.—Now what is this? You might say, "What a queer assumption, since nothing is perfectly rigid." What is the criterion for rigidity? What do we assume when we assume the parts are rigid?

Wisdom: If we put in the clause "assuming of course that the parts are rigid", aren't we explaining the part which rigidity plays in the calculus?

Wittgenstein: Well, but rigidity does not come into the calculus at all.

The point is that when we make a calculation with respect to a machine, the more rigid the parts, the more accurate the calculation. It is in the *application* that rigidity enters.

Suppose someone suggested that kinematics treats of perfectly rigid mechanisms. This is just like saying that the logical mechanism is perfectly rigid. But that does not mean we treat of any mechanism *which is rigid.*

My brain won't work, but I'll make a suggestion. Suppose we always explained the way in which something rotates by a hypothetical mechanism of this sort. Instead of giving the mathematical law of the way in which the velocity changes in terms of angular velocity etc., we give the mathematical law for the motion of an 'ideal piston' to which it is imagined as being joined. Every rotating motion would be described by a law of motion of a piston. This would be actually a logical machinery. And one might here say that the logical machinery is always infinitely rigid.

The question is: What is the criterion for the rigidity of a part? Is it that the mechanism moves in such-and-such a way? or is it something else? It may well be simply the movement of the wheel. For if we actually have a real piston and fly wheel connected by a rod, and we measure the velocity of the wheel and the way in which the piston moves,—then in certain circumstances we should say, "Yes, this rod is rigid." We should take a certain behaviour of the piston in connexion with a certain behaviour of the wheel as a criterion for the rigidity of the rod.

Perhaps it would help to take the example of a perfectly inexorable or infinitely hard law, which condemns a man to death.

A certain society condemns a man to death for a crime. But then a time comes when some judges condemn every person who has done so-and-so, but others let some go. One can then speak of an inexorable judge or a lenient judge. In a similar way, one may speak of an inexorable law or a lenient law, meaning that it fixes the penalty absolutely or it has loopholes. But one can also speak of an inexorable law in another sense. One may say that the law condemns him to death, whether or not the judges do so. And so one says that, even though the judge may be lenient, the law is always inexorable. Thus we have the idea of a kind of super-hardness.

How does this picture come into our minds? We first draw a parallel in the expressions used in speaking of the judge and in speaking of the law: we say "the judge condemns him" and also "the law condemns him". We then say of the law that it is inexorable—and then it seems as though the law were more inexorable than any judge—you cannot even imagine that the law should be lenient.[2]

2. Cf. *Remarks on the Foundations of Mathematics,* Part I, §118.

I want to show that the inexorability or absolute hardness of logic is of just this kind. It seems as if we had got hold of a hardness which we have never experienced.

In kinematics we talk of a connecting rod—not meaning a rod made of steel or brass or what-not. We use the word "connecting rod" in ordinary life, but in kinematics we use it in quite a different way, although we say roughly the same things about it as we say about the real rod: that it goes forward and back, rotates, etc. But then the real rod contracts and expands, we say. What are we to say of this rod: does it contract and expand?—And so we say it *can't*. But the truth is that there is no question of it contracting or expanding. It is a *picture* of a connecting rod, a symbol used in this symbolism for a connecting rod. And in this symbolism there is nothing which corresponds to a contraction or expansion of the connecting rod.

(Or: if we did talk of contraction and expansion of a rod in kinematics, we should mean something quite different—it would not be a matter of expansion produced by the application of heat.)

Thus if we say it has always the same length, we are led to suppose that it is very rigid, more rigid than anything which we meet in nature. We speak as if in kinematics we were dealing with connecting rods of a certain kind; that is to say, we speak of the difference between kinematics and a scientific description of a connecting rod as a difference between the objects dealt with by kinematics and by science.

What I wanted to talk of is logical inference and what one might call the peculiar rigidity or inexorability of it. I said something like "There is no such thing as a logical mechanism." I said this because I wanted to throw light on statements of this kind.

One might say, "Isn't this an absurd thing to say? For what is it whose existence you are denying?" It seems as though, if you deny it, you must know what it is.—Again and again, I'll either say such things, or we'll come across them. Compare: "There isn't such a thing as an infinitesimal."

When one says that there is no such thing as, for instance, a logical mechanism, one is making a fishy statement. At any rate, one's statement needs explanation. Part of what I wanted to do here was to show what sort of statement this is. I wanted to put us right about the idea of a logical mechanism—about the role which "mechanism" plays in logic.

Similarly, if I say that there is no such thing as the super-rigidity of logic, the real point is to explain where this idea of super-rigidity comes from—to show that the idea of *super-rigidity* does *not* come from the same source which the idea of *rigidity* comes from. The idea of rigidity comes from comparing things like butter and elastic with things like iron and steel. But the idea of super-rigidity comes from the interference of two pictures—like the idea of the super-inexorability of the law. First we have: "The law condemns", "The judge condemns". Then we are led by the parallel use of the pictures to a point where we are inclined to use a superlative. We have then to show the sources of this superlative, and that it doesn't come from the source the ordinary idea comes from.

XXI

How do we become convinced of a logical law?

We often say that we are convinced of the truth of logic, or of a particular logical law. But the difficulty is that when we normally say we are convinced of something, we can say what it would be like for us to be shown to be wrong or shown to be correct. But can we be shown to be right or wrong in logic? What would be the criterion?

(1) We might say: It is some very primitive kind of experience which corroborates logical laws.

(2) We say of a proof that it convinces us of a logical law.—But of course a proof starts somewhere. And the point is: What convinces us of the primitive propositions on which the proof is based? Here there is no proof.

If one thinks that it is certain experiences which convince us of the truth of logical laws, the point is to see what experiences these would be. And then one finds that one doesn't actually take any experience as corroborating a logical law. Take the law of contradiction. Suppose I said to someone, "Leave the room and don't leave the room", and he just stood there not knowing what to do. Would you say, "See, the law of contradiction works"? You would not take this experience as corroborating the law of contradiction.

In the same way, if someone tells me that there are two chairs in this room and two in that, and we count them and find that there are four chairs, we don't take this as a corroboration of $2 + 2 = 4$.

Or suppose we have '$(x) . fx . \supset . fa$'. Nobody would regard an experience as corroborating this. Which means that we don't use such a proposition as anything which is corroborated. That isn't the use we make of it—although it *might* possibly be.

I have read someone, an extremely intelligent man, who said that the law of identity is proved over and over again to us by experience, but we don't take the trouble to say every time, "This is identical with this." [1]—"This colour" [*Wittgenstein pointed to the wall*] "is identical with this colour." But suppose when I say "this colour" the second time, I find that the colour has changed. Should we say then that this tended to refute the law of identity? Obviously not.—The point is that "This colour is identical with this colour" has the jingle of a sentence, but it isn't used like "This wall is white, and that wall has the same colour", after which we look and find out that it has.

In the way in which laws of logic are not corroborated or invalidated by experience—the same applies to rules of deduction. Thus if we say that fa follows from $(x) . fx$, we do not regard any experience as showing either that it does or that it does not follow.

Compare saying that one thing follows from another with changing the unit of measurement—say, when you have a ruler marked in inches and fractions on one side and in centimetres on the other side. If you are given the inches, you can derive the

1. Perhaps Spencer, whose views are quoted and discussed by William James in *Principles of Psychology*, vol. II, ch. XXVIII ("Necessary Truths and the Effects of Experience").

measure in centimetres, and vice versa: you say, "It has thirty centimetres, therefore it has so-and-so many inches."—Why should one want to translate measurements in terms of centimetres? There may be various reasons. Say cloth is measured by the inch because people generally measure it with their thumbs. But somewhere else it is measured in centimetres, because they have price lists made up in some special way. So we may have reasons for changing the expression of measurement.

You might ask: What are we convinced of when we are convinced of the truth of a logical proposition? How do we become convinced of, say, the law of contradiction?

We first learn a certain technique of using words. Then the most natural continuation for us is to eliminate certain sentences which we don't use—like contradictions. This hangs together with certain other techniques.

Suppose I am a general and I receive reports from reconnaissance parties. One officer comes and says, "There are 30,000 enemy", and then another comes and says, "There are 40,000 enemy." Now what happens, or what might happen? I might say, "There are 30,000 soldiers and there are 40,000 soldiers"—and I might go on to behave quite rationally. I might, for instance, act as though there were 30,000, because I knew that one of the soldiers reporting was a liar or always exaggerated. But in fact I should of course say, "Well, one of you must have been wrong", and I might tell them to go back and look again.

The point is that if I get contradictory reports, then whether you think me rational or irrational depends upon what I do with the reports. If I react by saying, "Well, there are 30,000 and there are 40,000", you would say, "What on earth do you mean?" You might say, "Surely you can't imagine there being 30,000 *and* 40,000." But this could be answered in all sorts of ways. I might even draw a picture of it—for instance a blurred picture, or a picture of 30,000 here and of 40,000 there.

"Recognizing the law of contradiction" would come to: acting in a certain way which we call "rational".

Frege in his preface to the *Grundgesetze der Arithmetik* talks about the fact that logical propositions are not psychological propositions. That is, we cannot find out the truth of the propositions of logic by means of a psychological investigation—they do not depend on what we think. He asks: What should we say if we found people who made judgments contrary to our logical propositions? What should we say if we found people who did not recognize our logical laws *a priori,* but arrived at them by a lengthy process of induction? Or if we even found people who did not recognize our laws of logic at all and who made logical propositions opposite to ours? He says, "I should say 'Here we have a new kind of madness'—whereas the psychological logician could only say 'Here's a new kind of logic.' " [2]

This is queer. We wouldn't call a man mad who denied the law of contradiction—or would we?

Take this case: people buy firewood by the cubic foot. These people could learn a technique for calculating the price of wood. They stack the wood in parallelepipeds a foot high, measure the length and breadth of the parallelepiped, multiply, and take a shilling for each cubic foot.—This is one way of paying for wood. But people could also pay according to conditions of labour.

But suppose we found people who pile up wood into heaps which are not necessarily a foot high. They measure the length and breadth but not the height, multiply, and say, "The rule is to pay according to the product of length and breadth." Wouldn't this be queer? Would you say these people were asking the wrong price? Suppose that in order to show them what a stupid way of calculating the price of wood it is, I take a certain pile which they price at three shillings, and make it longer by making it less high. What if the heap piled differently amounted to £1—and they said, "Well, he's buying more now, so he must pay more."—We might call this a kind of logical madness. But there is nothing wrong with giving wood away. So what is wrong with this? We might say, "This is how they do it." [3]

Another case: Suppose someone wants to find out how many times 3 is contained in this lot of strokes: | | | | | | | | |. Then he may count this way:

2. Page xvi.
3. Cf. *Remarks on the Foundations of Mathematics,* Part I, §§142–152.

"Three, three, three, three—it goes *four* times."

That seems quite plausible. Suppose people even calculated this way when they wanted to distribute sticks. If nine sticks are to be distributed among three people, they start to distribute four to each. Then one can imagine various things happening. They may be greatly astonished when it doesn't work out. Or they may show no signs of astonishment at all. What should we then say? "We cannot understand them."

But—and this is an important point—how do we know that a phenomenon which we observe when we are observing human beings is what we ought to call a language? or what we should call calculating?

We most of us talk with the mouth—a few like me with the hands and mouth. And writing is ordinarily done with the hand. And so what we call a language is characterized not merely by its use but by certain other signs too; a criterion of people talking is that they make articulated noises. For instance, if you see me and Watson at the South Pole making noises at each other, everyone would say we were talking, not making music, etc.

Similarly if I see a person with a piece of paper making marks in a certain sort of way, I may say, "He is calculating", and I expect him to use it in a certain way. Now in the case of the people with the sticks, we say we can't understand these people—because we expect something which we don't find. (If someone came into the room with a bucket on his shoulders, I'd say, "That bucket must hide his head.")

We can now see why we should call those who have a different logic contradicting ours mad. The madness would be like this: (a) The people would do something which we'd call talking or writing. (b) There would be a close analogy between our talking and theirs, etc. (c) Then we would suddenly see an entire discrepancy between what we do and what they do—in such a way that the whole point of what they are doing seems to be lost, so that we would say, "What the hell's the point of doing this?"

But is there a *point* in everything we do? What is the point of our brushing our hair in the way we do? Or when watching the coronation of a king, one might ask, "What is the point of all this?" If you wish to give the point, you might tell the history of it.

What was the point of imitating gothic? It isn't clear in all that we do, what the point is.—But in the case of the people distributing the sticks, we would be struck by the pointlessness. Just as in the case where people calculate the price of wood in the queer way described.

Suppose I gave you a historical explanation of their behaviour: (a) These people don't live by selling wood, and so it does not matter much what they get for it. (b) A great king long ago told them to reckon the price of wood by measuring just two dimensions, keeping the height the same. (c) They have done so ever since, except that they later came not to worry about the height of the heaps. Then what is wrong? They do this. And they get along all right. What more do you want?

We are accustomed when we make experiments to record the results in a graph. And when the points lie like this, we know roughly what curve to draw:

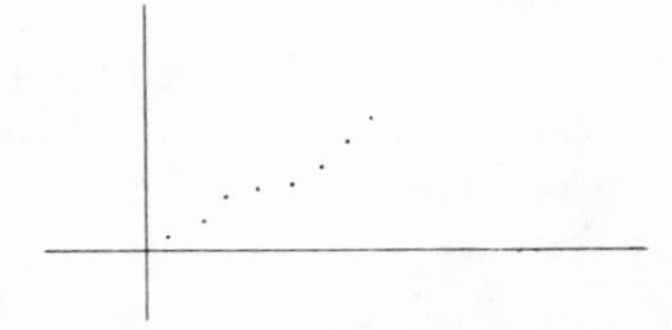

But then we may find that in economics someone draws a curve through the points even if they are distributed so:

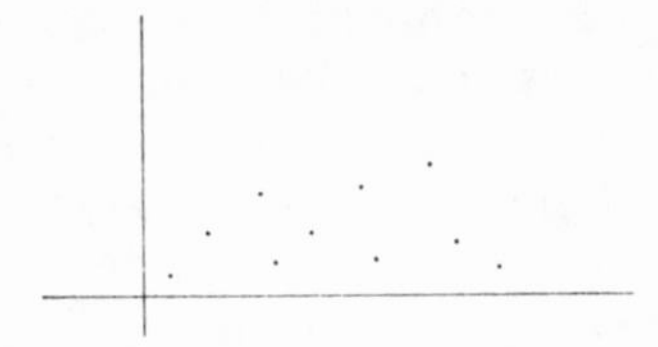

Here the practice degenerates into a ceremony. You might as well look into the entrails of a goose to predict something.—But why not? They say, after all, that it gives them some guidance.

A use of language has normally what we might call a *point*. This is immensely important. Although it's true this is a matter of degree, and we can't say just where it ends.

Suppose that in setting out the pieces for chess the kings are always used to determine who gets the white pieces. If at some time you and I used pawns for this, we would think it absurd if someone said, "This isn't chess." But suppose now that it is prescribed in the rules that one uses the kings. Would you call this 'not part of chess'? We would say, "It's not essential." We have, apart from any table of rules, an idea of the *point* of a game.—But what is regarded by one person as essential may be regarded by another as inessential; and it isn't always a clear issue.

The general who received the two contradictory reports, acted on them, and then won the battle—would *still* have acted in a queer way in our view. One would perhaps say, "What does he do with these reports? Perhaps he does not regard them as reports at all." We might call his use of the contradiction pointless or say that we don't understand it—though again it might be explained to us.

What's the conviction like, I asked, that the law of contradiction is true? Let's ask: What are the criteria for a person being convinced of a certain proposition?

(1) He says it in a tone of conviction.—But this isn't all.

(2) How he behaves, etc. I'd find out how he behaves before and after saying "I'm convinced that . . ." (for example, "I am convinced that this drink is poisonous"). If he says, "I am convinced that this drink is poisonous", and if he does not behave as if he wished to commit suicide, and if he then drinks it . . . we should not understand his statement.

How does one find out that a man is convinced of the law of contradiction? Well, he says "$(p). \sim(p. \sim p)$".—But how does one convince him of it?

You might say to him, "Now try—sit and don't sit." Though as a matter of fact, he does not then try to do anything, after a time he may well say, "No, I can't do any such thing." Or you might ask him, "Can you imagine it both raining and not raining?" Then what would actually happen might be that at first certain images come before his mind, and then no images come and he'd give up trying to imagine anything. Or in the case of "Sit and don't sit", he might consider various possibilities and reject them all. He might consider getting up, and then not getting up, and then might shrug his shoulders and say, "I can't do that."

Nothing happens; that is, he won't do anything. Why does he say, "I can't do that"? People do say this—although the case is so different [from other cases where we say "I can't"]. The analogy is: nothing happens.

Think of the fairy tale in which a prince wants a farmer's wife and so he sets the farmer various tasks. One of these is to fetch the prince a Klamank—which means nothing. The farmer sits down and cries, and a fairy asks him why. "The prince has told me to bring him a Klamank, and I can't do it." So the fairy gives the farmer a magic reed; and the farmer has only to touch a thing with the reed and it will follow him. So the farmer touches all sorts of things with it, and eventually he goes up to the prince with an enormous train behind him and says, "Here is a Klamank" —for it is something which the prince has never seen before and which might be called a Klamank.—The man who says "I can't both sit and not sit" is doing the same as the farmer when he said "I can't bring this." We make an analogy between: (a) an order which makes sense and which we can't obey, and (b) an order which sounds as if it makes sense but doesn't.

How do we get convinced of the law of contradiction?—In this way: We learn a certain practice, a technique of language; and then we are all inclined to do away with this form—on which we do not naturally act in any way, unless this particular form is explained afresh to us.

This has a queer consequence: that contradictions puzzle us. Think of the case of the Liar. It is very queer in a way that this should have puzzled anyone—much more extraordinary than you might think: that this should be the thing to worry human beings. Because the thing works like this: if a man says "I am lying" we say that it follows that he is not lying, from which it follows that he is lying and so on. Well, so what? You can go on like that until you are black in the face. Why not? It doesn't matter.

What does it mean to say that one proposition follows from another? One might say that it means "If we assert one proposition, we are then entitled to assert the proposition which follows." But what does *"entitled"* mean? Isn't one entitled to say anything?

Or you might say, "This is the technique." There is a certain use; and drawing the conclusion consists, say, in our writing so-and-so.—Take the ruler we mentioned before. We measure and say, "30 inches—and therefore 100 centimetres"; then we do so-and-so—and "therefore it weighs so much." That is how the technique of deducing one thing from another is used.

Now suppose a man says "I am lying" and I say "Therefore you are not, therefore you are, therefore you are not . . ." —What is wrong? Nothing. Except that it is of no use; it is just a useless language-game, and why should anybody be excited?

One might ask, "How on earth did this happen? How is it that we get a contradiction here although we do not usually get them?" In that case what is puzzling you is the lack of system. You want to know why a contradiction comes with "I am lying" and not with "I am eating".

In the first place, it doesn't happen in our ordinary use of "I'm lying". And if we have a use of "I'm lying" from which it follows "I'm not lying"—isn't this just a useless game?

Turing: What puzzles one is that one usually uses a contradiction as a criterion for having done something wrong. But in this case one cannot find anything done wrong.

Wittgenstein: Yes—and more: nothing has been done wrong. One may say, "This can only be explained by a theory of types." But what is there which needs to be explained?

Wisdom: It might be said that the theory of types decrees that one cannot make a statement about the statement one is making.

Wittgenstein: Cannot? But I do.

Wisdom: They would say that "I am lying" is not a statement about itself.

Wittgenstein: Ah, that is the point. We might ask, "Is it a sentence?" or "Is it a proposition?"—Tell me, what is the criterion for making a statement?

"If it is an entirely useless game, why did we ever think of playing it?"—Answer: Because "I am lying"—which is an ordinary statement—is analogous to "I am eating".[4] And then the point is to show cases in which we *would* use such a statement: for example, "He is 34—I'm lying, he's 32."

Wisdom: One might say that the theory of types shows that those who try to point out a different use of "I'm lying" do not succeed.

Wittgenstein: But "do not succeed"? Let's look into this.

Suppose I give a rule for the use of: "this ↗ ".

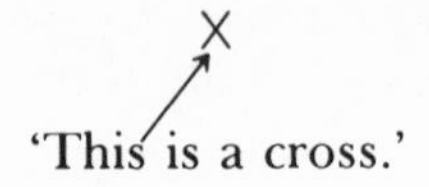

Then suppose I write:

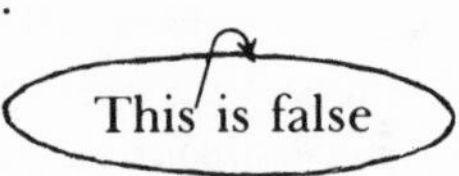

Is this a statement or isn't it? I'd say: I don't know; call it what you like.—How is it used? One way is:

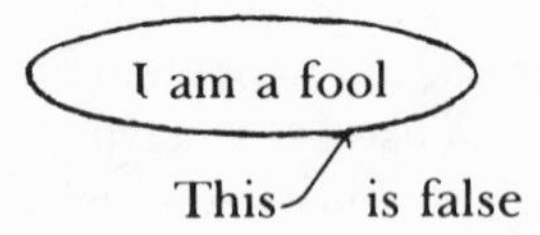

Here we'd know what to do, what follows from it, and so on. Whereas if we turn the arrow towards itself, we just wouldn't know what to do.

The words in the sequence "This is red" are what one calls an English sentence. But if I just write it on the blackboard, you may say this has no meaning at all—because there is no pointing gesture. And similarly if I write:

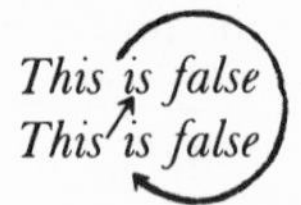

4. B and R give two different answers to the question "Why?", which have been combined.

This has no use either. (Suppose we sometimes had whole series of such sentences, where the statement only comes out at the end. Here we might go round and round in a circle for a quarter of an hour.)

If the question is whether this is a statement at all, I reply: You may say that it's not a statement. Or you may say it *is* a statement, but a useless one.

"The puzzle arises because one regards a contradiction as a sign that something is wrong."—There is a particular mathematical method, the method of *reductio ad absurdum*, which we might call "avoiding the contradiction". In this method one shows a contradiction and then shows the way from it. But this doesn't mean that a contradiction is a sort of devil.

One may say, "From a contradiction everything would follow." The reply to that is: Well then, don't draw any conclusions from a contradiction; make that a rule. You might put it: There is always time to deal with a contradiction when we get to it. When we get to it, shouldn't we simply say, "This is no use—and we won't draw any conclusions from it"?

Is Russell's logic vitiated by a contradiction?

Rhees: One might feel that by saying there is nothing wrong with a contradiction one is letting in the infection. For how are we to know that we must not allow other contradictions?

Wittgenstein: And why not?

Suppose that one uses Russell's logic in order to draw conclusions. Would this use be vitiated by the fact that a contradiction can be produced somewhere in Russell's logic? And how would it be vitiated? You've compared a contradiction to a germ; and that is the analogy which immediately springs to mind. One thinks of a doctor saying "You look all right from the outside, but this germ is a sign of your being fearfully ill inside." But then the question arises: What is the illness in this case?

What one is mainly afraid of is surely what is sometimes called a "*hidden* contradiction".—In what way "hidden"?

Now one can imagine an enormous number of rules or axioms written on an enormous blackboard. Somewhere I have said *p*, and somewhere else I said $\sim p$, and there were so many axioms I didn't notice there was a contradiction.

Or suppose that there is a contradiction in the statutes of a particular country. There might be a statute that on feast days the vice-president had to sit next to the president, and another statute that he had to sit between two ladies. This contradiction may remain unnoticed for some time, if he is constantly ill on feast-days. But one day a feast comes and he is not ill. Then what do we do? I may say, "We must get rid of this contradiction." All right, but does that vitiate what we did before? Not at all.

Or suppose that we always acted according to the first rule: he is always put next to the president, and we never notice the other rule. That is all right; the contradiction does not do any harm.

When a contradiction appears, then there is time to eliminate it. We may even put a ring round the second rule and say, "This is obsolete."

Suppose that we have a technique of finding hidden contradictions. For instance, suppose that we compare each rule with every other rule. Or in the case of logical systems, suppose that the axioms may be transformed so as to lead or not to lead to contradictions. Then there may be a technique for finding whether it will lead to contradictions: or there may be no such technique.

If there is no such technique, then it doesn't matter. It is not a case of *our* not having got it; the calculus simply has not got such a thing. If there is no technique, we ought not to talk of a hidden contradiction.

The word "hidden" has as many different meanings as there are methods of finding. When no method of finding has been laid down, there is no point in using the word "hidden".

Suppose we now use our rules, and one day we arrive at a contradiction. We may then say that we have not used the rules correctly; or we may want to change the rules.

I may give you the rules for moving chessmen without saying that you have to stop at the edge of the chessboard. If the case arises that a man wishes to make a piece jump off the chessboard, we can then say, "No, that is not allowed." But this does not mean that the rules were either false or incomplete.—Remember what was said about counting. Just as one has freedom to continue counting as one likes, so one can interpret the rule in such a way that one may jump off the board or in such a way that one may not.

But it is vitally important to see that a contradiction is not a germ which shows general illness.

Turing: There is a difference between the chess case and the counting case. For in the chess case, the teacher would not jump off the board but the pupil might, whereas in the counting case we should all agree.

Wittgenstein: Yes, but where will the harm come?

Turing: The real harm will not come in unless there is an application, in which case a bridge may fall down or something of that sort.

Wittgenstein: Ah, now this idea of a bridge falling down if there is a contradiction is of immense importance. But I am too stupid to begin it now; so I will go into it next time.

XXII

It was suggested last time that the danger with a contradiction in logic or mathematics is in the application. Turing suggested that a bridge might collapse.

Now it does not sound quite right to say that a bridge might fall down because of a contradiction. We have an idea of the sort of mistake which would lead to a bridge falling.

(a) We've got hold of a wrong natural law—a wrong coefficient.

(b) There has been a mistake in calculation—someone has multiplied wrongly.

The first case obviously has nothing to do with having a contradiction; and the second is not quite clear.

Whatever example one constructs will seem extremely crude. But that does not matter here.—Imagine that some great man taught human beings to multiply, divide, etc. They were very slow in learning, clumsy, etc. Shortly before he died, he left them one more mathematical proposition, namely

$$3678 \times 19375 \neq -----$$

which is in fact wrong. Then later they find that there is a contradiction. The master had left them all sorts of rules, but he left *one* rule which didn't work.

How would this affect a practical problem? Would it affect it at all? We might say that when the problem first arose (for example, how many soldiers there were, when they would have to calculate the product of these numbers) they would not know what to say—whether to say what the master had said or something else.

What would they in fact say? They might say that the master was wrong and abolish the rule. But must they? Couldn't they say, "Now we'll assume both this and the opposite"—and now the question is how they will use it. Or they could say, "The master was right, but when we count, one soldier vanishes, or comes into existence."

I am not recommending this kind of arithmetic. All I mean is: the mere fact of a contradiction would not necessarily get them into any trouble.

What they do when they get to the contradiction will depend on what reasons they had for holding to that formula—in a sense, on how much that formula means to them. I made up a very silly example because I couldn't think of any reasons they could have.

Turing: The sort of case which I had in mind was the case where you have a logical system, a system of calculations, which you use in order to build bridges. You give this system to your clerks and they build a bridge with it and the bridge falls down. You then find a contradiction in the system.—Or suppose that one had two systems, one of which has always in the past been used satisfactorily for building bridges. Then the other system is used and the bridge falls down. When the two systems are then compared, it is found that the results which they give do not agree.

Wittgenstein: Now look. Suppose I am a general and I give orders to two people. Suppose I tell Rhees to be at Trumpington at 3:00 and at Grantchester at 3:30, and I tell Turing to be at Grantchester at 3:00 and to be at Grantchester at the same time as Rhees. Then these two compare their orders and they find "That's quite impossible: we can't be there at the same time." They might say the general has given contradictory orders.

This simply means that given a certain training, if I give you a contradiction (which I need not notice myself) you don't know what to do. This means that if I give you orders I must do my best to avoid contradictions; though it may be that what I wanted was to puzzle you or to make you lose time or something of that sort.

That is one thing: (a) We do in fact try to avoid contradictions. (b) Unless we wish to produce confusion (given our training) we *have* to avoid contradictions.—But it is an entirely different thing to say that we ought to avoid contradictions in *logic.*

If we talk of logic, we think of the calculations and ways of thinking which we do in fact have—the technique of language which we all know. And in this technique contradictions don't normally occur—or at least occur in such restricted fields (e.g., the Liar) that we may say: If that's logic, it doesn't contain any contradictions worth talking about. We as a matter of fact avoid contradictions and are even inclined to call it illogical thinking if there are contradictions. But you might say: This is only one logic, and in others you may have as many contradictions as you like.

A contradiction is, say, an expression of the form '$p . \sim p$'. At least, I say that, and it sounds all right, but in a sense it is bosh. Because 'p and not p' is English; we know what 'and' and 'not' mean, and 'p' stands for propositions like "it rains" which we all know.—But in '$p . \sim p$': who says this—'$\sim$'—is a negation? This is a curl and this is a dot. What makes this a sign of negation, this a logical product, and so on? I can see two things: either it's the use *in* the calculus or it's the use *outside* the calculus.

That we ought not to get '$p . \sim p$' comes from our thinking of [the signs] with the normal application—because this is the way we actually calculate.[1] If we had a calculus in which 'p', we said, stood for propositions, and in which '$\sim$' and '.' are used in a way similar to the use of 'not' and 'and', and if we then allowed '$p . \sim p$'—then if you like you could say that new rules have now been given for those symbols and that what looked like a logical product or a sign of negation was not really so.

'$p . \sim p$' in this calculus might stand for "Jack and not-Jack".

1. There are two versions of this sentence, one inaccurate, the other sketchy and incomplete.

Compare the use of "Jack and not-Jack is in this room", meaning "Jack and also some others are in this room."—You might say that this is cheating because "Jack" is not a proposition. But "Jack" may be used as a proposition—for example, "Come here, Jack" or "Jack is here".

What I mean is this. If you ask, "What would happen if we had another logic? What would go wrong?" I would say, "Nothing, except that we might not be inclined to call it logic any more." Think of the case where people have a queer way of calculating a price for the wood: we might not be inclined to call it calculation at all.

It's like this. If we do, say, physics, or if we do zoology and give an account of an animal, we don't want contradictions in that account. If we then think of mathematics or logic as a sort of physics—compare Frege's view that a law of logic is a law in terms of which we have to think in order to think what is true (similar to laws of physics, but completely general) [2]—if we think like this, we think at once: "Then there mustn't be any contradictions in logic." But this is queer. For that there must be no contradictions in logic ought itself to be a logical law.

You may want to say, "Logic and mathematics can't reveal any truths if there are contradictions in it." But Russell as a matter of fact makes every proposition in it a tautology, which is just as bad. And he might just as well have made them all contradictions; for we have seen that we could do all logic with contradictions.

This not having contradictions characterizes a peculiar technique of ours.

You might say that if you had contradictions, your calculus would be useless. But this would depend on what kind of use you wanted to make of it.—One wants to say, "You couldn't make the same use of it (arithmetic, say) which we make now, if it contained contradictions."

But: "use of *it*"? This is queer. As a matter of fact we use an arithmetic which has no contradictions. Now if we had a different

2. *Grundgesetze,* I, xv.

arithmetic, whether we could or couldn't use it in the same way depends on whether we would still call it "using it in the same way". We might not be willing to call anything the *same use.*

Suppose we have:

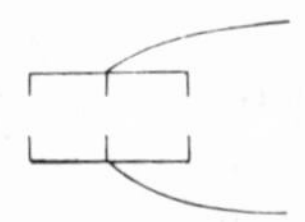

This shows two electrodes; you press the top one down, and when it makes contact, three bells ring.

"If this one had two prongs, then we could not use it in the same way."—There is something queer about this. Is it an arithmetical or an experiential statement? "If the electrode were made of copper instead of iron we could not use it in the same way because the resistance would be greater"—this is a statement of physics. But that we could not use it in the same way if this had two prongs—we are inclined to say that this is simply a matter of arithmetic.

The point is: What is 'the same way'? If it is a statement of physics, then "it can't be used in the same way" means it can't do the same things—for example, the three bells would not ring. But what if they do ring? Then it seems as though we can use the two-pronged electrode in the same way after all.

But there is another way of taking it. And the difficulty is this. If "one way" is characterized by this figure:

and "another way" by this:

then it is a matter of definition that we cannot use it in the same way.

Similarly with arithmetic. "We could not use another arithmetic in the same way." Do you mean that we could not use it to build houses? Well, we'll see; this is experiential. Or do you mean that we don't call this "the same way"?—The trouble is to distinguish between what already lies in the *picture* of an arithmetic, and what does not lie in that picture.

Prince: Could we take this example: Suppose we have two ways of multiplying which lead to different results, only we don't notice it. Then we work out the weight of a load by one of these ways and the strength of a brass rod by the other. We come to the conclusion that the rod will not give away; and then we find that in fact it does give way.

Wittgenstein: This comes to the same as having an arithmetic in which the associative law (the law that $a \times (b \times c) = (a \times b) \times c$) does not hold. Then in calculating the volume of this book we shall get different results according to how we multiply the length and the breadth and the height.—But this does not help us. We might take one answer as the right one, or we might do anything.

Turing: We tried to find why people were afraid of contradictions, and we talked last time of hidden contradictions. This example of Prince's shows that practical things may go wrong if you have not seen the contradiction.

Wittgenstein: By "seeing the contradiction" do you mean "seeing that the two ways of multiplying lead to different results"?

Turing: Yes.

Wittgenstein: The trouble with this example is that there is no contradiction in it at all. If you have two different ways of multiplying, why call them both multiplying? Why not call one multiplying and the other dividing, or one multiplying-A and the other multiplying-B, or any damn thing? It is simply that you have two different kinds of calculations and you have not noticed that they give different results.

Turing: Might it not be an axiom that the two should give the same result?

Wittgenstein: Yes—[just as] you might take Fermat's law as an axiom.

It might be that if it were taken as an axiom, then you would not know what would happen if a contradiction were discovered. Of course, if we just took it as an axiom for fun, we can imagine discarding the axiom immediately we discovered the contradiction. But if we really had a reason for taking it as an axiom—for instance, if the master had left it to us—then we need not give it up.

Prince has talked of our not noticing that two kinds of multiplication give divergent results. But what if we *never* noticed the divergence? Is it necessary that something should go wrong with the brass rod? Might it not always be perfectly all right?

How do you know you have not left out a number when you count? We might under certain circumstances say [we had left out a number]—if we were very tired and added with a different result every time. As a matter of fact this very seldom happens.

It is difficult to imagine we hadn't noticed the contradiction at all—this is important. But suppose we haven't noticed it and suppose that nothing goes wrong: the bridge doesn't fall or the brass rod doesn't break. Is our calculation wrong? I'd say: Not at all. We've done everything perfectly all right. Perhaps at a later stage we might say that the brass rod constantly changes its elasticity or something of that sort.

The question is: Why are people afraid of contradictions? It is easy to understand why they should be afraid of contradictions in orders, descriptions, etc., *outside* mathematics. The question is: Why should they be afraid of contradictions inside mathematics? Turing says, "Because something may go wrong with the application." But nothing need go wrong. And if something does go wrong—if the bridge breaks down—then your mistake was of the kind of using a wrong natural law.

Is Prince's case a case of a "hidden contradiction"? And if something is a "hidden contradiction", does it do any harm while it is—as you might say—hidden?

You might say that with an open contradiction we would not know what to do; we would not know what use to make of it. And what about a "hidden contradiction"? Is it there as long as it is hidden?

Turing: You cannot be confident about applying your calculus until you know that there is no hidden contradiction in it.

Wittgenstein: There seems to me to be an enormous mistake there. For your calculus gives certain results, and you want the bridge not to break down. I'd say things can go wrong in only two ways: either the bridge breaks down or you have made a mistake in your calculation—for example, you multiplied wrongly. But you seeem to think that there may be a third thing wrong: the calculus is wrong.

Turing: No. What I object to is the bridge falling down.

Wittgenstein: But how do you know that it will fall down? Isn't that a question of physics? It may be that if one throws dice in order to calculate the construction of the bridge it will never fall down.

Turing: If one takes Frege's symbolism and gives someone the technique of multiplying in it, then by using a Russell paradox he could get a wrong multiplication.

Wittgenstein: This would come to doing something which we would not call multiplying. You give him a rule for multiplying; and when he gets to a certain point he can go in either of two ways, one of which leads him all wrong.

Suppose I convince Rhees of the paradox of the Liar, and he says, "I lie, therefore I do not lie, therefore I lie and I do not lie, therefore we have a contradiction, therefore $2 \times 2 = 369$." Well, we should not call this "multiplication"; that is all.

It is as if you give him rules for multiplying which lead to different results—say, in which $a \times b \neq b \times a$. That is quite possible. You have given him this rule. Well, what of it? Are we to say that you have given him the wrong calculus?

Turing: Although you do not know that the bridge will fall if there are no contradictions, yet it is almost certain that if there are contradictions it will go wrong somewhere.

Wittgenstein: But nothing has ever gone wrong that way yet. And why has it not?

A person who doesn't think about it much might imagine that $5^{(6^4)}$ is the same as $(5^6)^4$. He calculates sometimes one way, sometimes the other, and doesn't notice that he gets different results. This again is parallel to thinking that $\sqrt{25+36} = 5 + 6$.

Suppose mathematicians of a certain period thought the root of a sum was the sum of the roots.—But what is it we are to assume? Are we to assume they never bothered to compare the results? We can imagine them learning a technique and teaching it in their schools—and then after a time saying, "Oh, it no longer works, because the two give different results."—But what should we call the hidden contradiction? Where would it be hidden? And when is it hidden and when does it cease to be?

Is it hidden as long as it hasn't been *noticed?* Then as long as it's hidden, I say that it's as good as gold. And when it comes out in the open it can do no harm.

[*To Turing*] Before we stop, could you say whether you really think that it is the contradiction which gets you into trouble—the contradiction in logic? Or do you see that it is something quite different?—I don't say that a contradiction may not get you into trouble. Of course it may.

Turing: I think that with the ordinary kind of rules which one uses in logic, if one can get into contradictions, then one can get into trouble.

Wittgenstein: But does this mean that with contradictions one *must* get into trouble?

Or do you mean the contradiction may tempt one into trouble? As a matter of fact it doesn't. No one has ever yet got into trouble from a contradiction in logic. [It is] not like saying "I am sure that that child will be run over; it never looks before it crosses the road."

If a contradiction may lead you into trouble, so may anything. It is no more likely to do so than anything else.

Turing: You seem to be saying that if one uses a little common sense, one will not get into trouble.

Wittgenstein: No, that is *NOT* what I mean at all.—The trouble described is something you get into if you apply the calculation in a way that leads to something breaking. This you can do with *any* calculation, contradiction or no contradiction.

What is the criterion for a contradiction *leading* you into trouble? Is it specially *liable* to lead you into trouble?

It cannot be a question of common sense; unless *physics* is a question of common sense. If you do the right thing by physics, physics will not let you down and the bridge will not collapse.

You might say, "If we applied Frege's calculus using the Russell paradox, this would mean simply that we had multiplied wrongly."—Or you might say, "Frege does not teach us to multiply, because if we go through Russell's paradox we can get to anything." You can say, "Frege allows a wrong turn through which we can get to any result at all. Give a man Frege's *Grundgesetze* and he can get anything."

But what if you do say this? What then?

Turing: If you say that contradictions will not really lead one into trouble, you seem to mean that one will take up towards contradictions the attitude which I described.

Wittgenstein: You might get $p. \sim p$ by means of Frege's system. If you can draw any conclusion you like from it, then that, as far as I can see, is all the trouble you can get into. And I would say, "Well then, just don't draw any conclusions from a contradiction."

Turing: But that would not be enough. For if one made that rule, one could get round it and get any conclusion which one liked without actually going through the contradiction.

Wittgenstein: Well, we must continue this discussion next time.

XXIII

We were in a mess at the end of last time and we shall probably get into the same mess again today. I find it very difficult to go on from the point I reached a short while ago; but I must go on from that point. Philosophy is like unravelling a ball of wool. It's no use *pulling* at it. And I am apt to pull.

We talked about how a contradiction might be harmful. Let's take one or two examples of this.

Prince talked about two kinds of multiplication. In a way there are two kinds of multiplication, for example, the proof that $4 = 5$, which puzzles small children. It is based on proving that $4 \times 0 = 5 \times 0$, and then dividing both sides by 0, or using the rule: if $a \times b = c$ and $d \times b = c$ then $a = d$.

Suppose we have this case: a man could be told all the rules for multiplication—only he was not told that you must not cancel a 0. And so he could through this kind of thing get to any result. It is conceivable that in this way you might give a person a set of rules without being aware that you have given him a rule which you haven't properly cut down, and which allows *any* conclusion—which you didn't want to allow. He might try to check his results by these means and always find them right.

Consider another example. Suppose people had built a prison, and that the point of it is to keep the prisoners apart. Each prisoner can move along certain corridors and into certain rooms; but the rooms and corridors are so arranged that no two prisoners can ever meet.

We could imagine that the system of corridors is very complicated—so that you might not notice that one of the prisoners can after all get by a rather complicated route into the room of another prisoner. So you have forfeited the point of this arrangement.

Now suppose first that none of the prisoners ever noticed this possibility, and that none of them ever went that way. We could imagine that whenever two corridors cross at right angles, they always go straight on and never think of turning the corner. And suppose that the builder himself had never been struck by the possibility of their turning the corner at a crossing. And so the prison functions as good as gold.

Then suppose someone later on finds this possibility and teaches the prisoners to turn the corner. Can we say, "There was always something wrong with this prison?"

Well, we can say several different things: (1) the prison functioned all right; (2) we can say that it *was* wrong, in the sense that one day people found this way, and that perhaps things went wrong and the prison became useless.

(Here I've made an obvious system: "Whenever two corridors

cross at right angles, etc." But it may not be as simple as that, and the result may still be that they actually never went that way.)

Let's go back to the contradiction with multiplication by 0. Suppose we had neglected to tell him that he must not multiply in this way. If we had not told him that he cannot always say $ab = c$ if $ab\alpha = c\alpha$, he might get wild results which we don't want. —In this sense, if we had a calculus in which a man was liable to go wrong—if he went by way of a contradiction to some absurd thing, or checked some absurd result by seeing whether it agreed with this calculation—then we should perhaps say we had neglected to make the rules stringent enough.

I have two things to say about this. The first is that the contradiction itself need not be called false at all. And if the danger is simply that someone might go this way unawares and get absurd results which we do not want, then the only thing is to show him which way not to proceed from a contradiction.

Take Russell's contradiction: There are concepts which we call predicates—"Man", "chair", and "wolf" are predicates, but "Jack" and "John" are not. Some predicates apply to themselves and others don't. For instance "chair" is not a chair, "wolf" is not a wolf, but "predicate" is a predicate.

You might say this is bosh. And in a sense it is. No one says " 'Wolf' isn't a wolf." We don't know what it means. Is "Wolf" a name?—in that case Wolf *may* be a wolf. If someone asked, "Is 'wolf' a wolf?", we simply would not know what to answer.

But there is one way in which Russell would have used it. Nobody would say, " 'Wolf' is a wolf", but " 'Predicate' is a predicate" people would say. We can distinguish between predicates which apply to themselves and those which don't, and form the predicate "predicate which does not apply to itself". Does this apply to itself or not? It is clear that if it does apply to itself, then it does not; and that if it does not, then it does. From this it presumably follows that it both does and does not apply to itself.

I would say, "And why not?" If I were taught as a child that this is what I ought to say, I'd gladly say so.

What is queer about this sentence is that we don't know what on earth to do with it, any more than we know what to do with " 'Wolf' is a wolf."

I don't say " 'Wolf' is a wolf" has no meaning. I don't know how to decide this. But I will say it hasn't a use—although under certain circumstances (when "Wolf" is a name, say) it may. We don't distinguish between having a meaning and not having a meaning, but between being used and not being used. This is very important when, for example, the question arises of whether mathematics is just a game with symbols or whether it depends on the meanings of its signs. This question vanishes when one ceases to think of meaning as being something in the mind. If you say, "The sign '2' has no meaning", do you want to say we don't count chairs? or do you just want to distinguish between mathematics and its application? There is no question of giving it a meaning apart from an application.

And now about contradictions. Whether we're to say they have a meaning I don't know—but it's clear they don't have a use. The point is: Don't think of a contradiction as a 'wrong proposition' ("Surely this isn't so" etc.). But this doesn't mean that a contradiction can't be pernicious, if it actually misleads us.

"With a little common sense you won't fall into the trap—you won't go via a contradiction." I said a short time ago that I didn't want to say that; and that's true. But I wanted to say something rather similar.

How can common sense stop you from going this way? For if it can, what does it provide?

It is common sense not to be afraid that the engine driver may just overlook Cambridge and drive on.—But one wouldn't call this [which we're now concerned with] common sense.

The point is whether there is or isn't something which prevents us from using the calculus like that. Suppose it were our education or training which prevented us—then that would be all right.

I must be very careful here. I am at a dangerous point and am likely to fall into the trap of meddling with the mathematicians.

Consider Russell's contradiction, and suppose that it had never been found. Should we say that on account of this, the foundations of mathematics would have been *wrong?* [. . .] [1]

Turing: Surely one can at any rate say that we have got now to build a new prison; and one ticks off the architect and tells him to look at the plans of the new prison very carefully before building it.

Wittgenstein: I agree entirely. But there are two points which are not clear.

We agree that the point of avoiding a contradiction is not to avoid a peculiar untruth about logical matters but to avoid the ambiguity that results—to avoid getting to that place from which you can go in every direction. A contradiction might forfeit the point of our calculus. So we scrutinize the logical calculus beforehand, just as Turing says we scrutinize the plans of the prison.

But there are two points here. First, you may or you may not know what is meant by this sort of scrutiny.—Suppose that in the prison there are air ducts and no one had ever thought of people going through an air duct. But then someone does get through an air duct. We might say to the architect, "Trace every air duct."

Suppose one called the air ducts a hidden way of escape, and now we said, "Trace every hidden way of escape." This might mean "Trace every air duct—and do it systematically." He now knows what to look for; and "Trace every air duct" gives a method of searching.

But suppose you said, "Search every hidden way of escape" and then, when he had traced every air duct and corridor, he said, "Is there anything else? Perhaps a prisoner might contract and get through the water pipes."—Then "Search every hidden way of escape" is quite different.

There are two cases. (1) I have a method of finding a contradiction, and then I can say it's *hidden* (in the sense in which the product of 18 X 28 is hidden as long as I have not calculated it). (2) We are vague about it. We are on the lookout for contradictions in systems.

One might say, "Russell's contradiction has put us on our

1. Wittgenstein probably asked another question, of which there is no record.

guard. A contradiction may lurk anywhere." To which we might reply, "Don't be so nervous. You're being silly." "Hidden" means: hidden in this way or that way.—Compare the case of a man who says, "An enemy may be hidden in this room." He may then search the room; he doesn't mean the enemy may have contracted into an air particle. If he supposes the enemy has turned into a sofa and may pounce out on him at any moment, that's no longer what we'd call hidden.

There is the case in which you have a calculus and later find a contradiction in it. We might also say that as soon as you've found the contradiction, it is no longer the same calculus. That is why I gave the example of the corridors.

This hangs together with the question: In what way can you say you find out something *new about* a calculus—as opposed to *adding* something to the calculus?

There is a difference, according to whether we want to talk about first principles—about the "foundations of mathematics"; or whether we want to talk about a particular calculus.

Suppose for some reason someone suspected a danger in a particular case. He would either already *have* a method for finding out whether there was such a thing or not; or he might have investigated by this method, and yet say, "Perhaps there is a contradiction still"—*now being entirely vague as to what he had overlooked.*

Take the case of the architect who has traced the air ducts and then says, "Maybe there is some other means of escape." He is then indefinite and has no method of checking up, but will sort of grope about. One cannot blame him, although one can say, "Don't be hysterical."—If you have no idea at all what you are looking for, then there is no clear limit where we'd say you should stop. Something may turn up any day.

The same applies to the calculus. If you are thinking of a particular *way* in which a contradiction may arise, you may, for example, go through the rules and check them in this respect. But if you are not, you may still grope about, and you may even find a contradiction in this way.—But then we must say that any rule may be reinterpreted—reinterpreted naturally or unnaturally. And if you interpret it in some new way, a contradiction may arise.

Turing: But in practice the question of rules being reinterpreted does not come in seriously.

Wittgenstein: This is very important.—Given a set of axioms, there may or may not have been provided a method for seeing whether there is a contradiction hidden in them. For instance, in Frege's system one might try all the possible ways in which the rules can be combined, although that would be tedious.

One may have no method for finding contradictions—what is one to say then to the question "Are there any contradictions in this calculus?" This is why I gave the example of the corridors. I said that no one had ever thought of turning the corner. Now do you know what you have not thought of?

"I've thought of everything."

Can you stop a man looking for a way to make the right hand and the left hand coincide—if he says just that he has not yet found a way? If you say, "You see, this doesn't work", . . . he says, "I know; I haven't found it."—We simply *decide* that there isn't a way.

The people who went this way and that way in the prison said they had explored all the avenues. And when someone taught them to turn the corner, they said, "Have we been blind all the time?" Why couldn't this happen to us—in the case of the two hands? This is vastly important. We do not imagine a case of reinterpreting a rule, just as the prisoners did not imagine anyone turning the corner.

What would make them turn the corner? Well, it might be that originally they had only right-angled crossings $+$ and very narrow forks Y ; and that then they got crossings which were halfway between a narrow fork and a right-angled crossing. Then it might seem most natural to them to turn the corner. Similarly, by surrounding $\sqrt{-1}$ by talk about vectors, it sounds quite natural to talk of a thing whose square is -1. That which at first seemed out of the question, if you surround it by the right kind of intermediate cases, becomes the most natural thing possible.

I want to make [clear the] difference between (1) what sort of methods are adopted in mathematics in particular cases, (2) the question of what methods are *fundamental,* and if they are not adopted then the whole of mathematics rocks.—I'm trying to get at the notion that without certain methods, it isn't mathematics, or is wrong in principle.

If you take Frege's logic as a calculus dealing with "and", "not", and so on,—it is all right as it is. It is rather tedious and hardly ever used; but if for some reason or other you have to avoid a contradiction in it, I hope you will.—*But this has nothing to do with the idea of giving foundations for mathematics.* It is dealing with a particular calculus. There is one reason *why* this calculus seems to underlie mathematics: these words—"if", "and", "equals", . . .—come in at any moment, and then these rules are applied.

If you based something on this system, I don't see that it would necessarily be detrimental if there were a contradiction in it, as long as this contradiction is just not used as a thoroughfare or circus. Then this calculus fulfils its particular purpose.—This calculus might be used (a) to base something on (as Frege does), or (b) to calculate with (as nobody ever does). So we might say: Everything in the calculus works all right as long as we do not pass through the contradiction.

Why should you say even when the contradiction is discovered, "Now everything is wrong"? Not even the contradiction itself is wrong, or a false mathematical proposition. The only point would be: how to avoid *going through* the contradiction unawares.

Turing: Suppose that one has a circus off one of the main corridors. The prisoners, when they find that they can turn right and get into this circus, then find that they can also turn right at other places and thus get into all sorts of places where they were not intended to go. It is the turning right that is the trouble; the circus is only a symptom. And one cannot get rid of the trouble simply by barring the circus.

Wittgenstein: Yes; but it does not apply to Frege's logic—in the sense that Frege's logic is all right: we can go the way that Frege went. From every point we might go the wrong way and get into a contradiction; but as a matter of fact we do not. And if we suppose that he actually produces the foundations of arithmetic, there is no further trouble—there are no laws which go wrong.

It isn't actually that people went through doors into places from which they could go any damn where. It isn't true that this happened with Frege's logic. If they did this, Frege's logic would be no good, would provide no guide. But it does provide a guide. People don't get into those troubles.

Turing: You said that the point of Frege's logic was as a basis for arithmetic.

Wittgenstein: Yes; and that is very vague. We must ask later what is meant by it.

Turing: Yes; but if he uses it for that purpose, and if he takes the line he does, knowing beforehand the things which are in arithmetic; and if he just doesn't pay attention to other ways he might have gone—if I see he could have used it to prove anything, and not just arithmetic, then I don't feel very impressed by it as a basis for arithmetic.

Wittgenstein: I should say the same.—The point I'm driving at is that Frege and Russell's logic is not the basis for arithmetic anyway—contradiction or no contradiction.

If it is impossible to close the doors to the circus—if we just don't know which things to eliminate and which not—then the calculus is no use to calculate with. It is a case of something which isn't a calculus at all.

Suppose I were led through corridors by voices, and not by the corridors themselves at all. The corridors leave any damn thing open and I can go anywhere. That would be like saying "He goes through the corridors of logic, led by arithmetic." This is not entirely true of Frege's logic. He was led by the normal rules of logic: the rules of such words as "and", "not", "implies", and so on. He was led also by our normal use of words. As we never ask whether "Fox" is a fox or "Predicate" is a predicate, the question didn't arise and he never got into trouble.—He went partly according to the way we go in ordinary English (or German) and partly according to the way we go in mathematics. He got $\phi(a)$ from mathematics; and the new thing about the symbolism was making this into a predicate. [But he didn't get to $\phi(\phi)$, so that the contradiction never came up. Here] he was led by something, not entirely by arithmetic; for if we express $\phi(\phi)$ in ordinary words, we don't know what we're talking about.[2]

So it is not quite right to say "Frege might have proved anything else." And this is shown by the fact that Russell, almost immediately on finding the contradiction, found a remedy in the theory of types: "We would never say 'Fox' is a fox; so eliminate that."

On the other hand, would you say Frege's calculus—apart from the question whether anything can be based on it—was worthless? I suppose not. You could imagine a very good use for it, for these long formulae with "if", "and", etc.

Turing: I could not find a use for it without modifying it. For instance, one would have to say, "Avoid expressions like '$\phi(\phi)$' " and a good deal of that sort of thing.

Wittgenstein: But is this a modification? You can call it a modification if you like, but it is just a question of closing certain doors. You would only be wary in certain special cases—there is a vast field in which the formulae would be all right.

"It would be all right with a little common sense"—there is some truth in that. It would be all right if you forgot it was intended as a basis for mathematics and simply used it for drawing rather complicated conclusions. It would be a dull calculus; but the contradiction doesn't vitiate it for that purpose.

Turing: If one eliminated the contradiction, then it would be all right. But if one simply avoids what feels fishy, then I would say that the contradiction did vitiate it.

Wittgenstein: What I want to talk about is: (1) In what sense can one say at all that logic is the foundation of arithmetic? (2) In what sense can one talk about "the truths of logic"? My point in talking about contradiction was to show what sort of thing would happen if you neglected the truths of logic. This was to find out in what sense we can say that they are *a priori*—the rock on which the whole thing rests.

2. The four accounts of the paragraph are quite different. Only B and R have any record of the second half; they are sketchy and differ in sense. That part of the paragraph is conjectural: it is *a* way of making sense of the pieces.

We have seen that if we didn't recognize a contradiction, or if we allowed a contradiction but, for example, did not draw any further conclusions from it, we could not then say we must come into conflict with any facts.—You *could* say that if we allow a contradiction, in the sense that we allow anything to follow from it, then we have given up any idea of a calculus at all.

If we "go against logical laws", we don't come into conflict with any facts; but if we don't recognize the law of contradiction *in this sense,* then not recognizing the law of contradiction means not calculating at all. Or we might say, "This is a calculus, but quite a useless one." So that in *this* sense, recognizing or not recognizing the truths of logic . . . and what does it mean: to recognize or not to recognize the truths of logic? To swear by logic? To say "Surely this cannot be so" in a convinced tone of voice? Or just not to do a certain thing, for example, not to produce an "utterly useless calculus"? I'd say: Yes, if anything, it's the latter. *It isn't that we are convinced of a particular truth. But rather that we want to do so-and-so.* Going against logic means *doing* something we don't want to do: not-calculating as opposed to calculating, or not-drawing a conclusion as opposed to drawing a conclusion.

Next time I hope to start with the statement: "The laws of logic are laws of thought." The question is whether we should say we cannot think except according to them, that is, whether they are psychological laws—or, as Frege thought, laws of nature. He compared them with laws of natural science (physics), which we must obey in order to think correctly.[3] I want to say they are neither.

One can say: "If you do this, you aren't doing what we'd call thinking", or "If you do this, we wouldn't call it calculating."

Is

$$\begin{array}{r} 4 \\ 5 \\ \hline 16 \end{array}$$

—where you can write down anything which comes into your head for 16—an arithmetical operation? Call it what you like.

3. *Grundgesetze,* I, xv

—It misses out some of the most essential points of a calculus. So you may say that logic gives you some of the most important characteristics of what's called thinking, calculating, etc. And there is no clearcut distinction; there are things which we should not know whether or not to call calculating or thinking.

XXIV

I wanted to go on with the notion that the laws of logic are laws of thought. I've been trying to explain what would go wrong if we assumed that the law of contradiction, for example, doesn't hold.

"The chair is not both brown and not brown."—This is a sentence we arrive at by continuing a certain line of the practice of using sentences. That is to say, if we gave the sentence "It's brown and not brown" a meaning, we would feel that we had broken the line of our ordinary practice of language.

So in this sense, the law of contradiction was just an intersection point of certain lines if drawn long enough. The logical laws in this sense represent a particular way of using sentences and transforming them. I mean: if you now ask, "What would go wrong if we didn't recognize the law of contradiction to be true?" the answer would be: it would not then be the calculus we want; we would not then make the transformations we do make, use propositions as we do use them.

For one thing is clear. The law of contradiction is a result of continuing in a particular way the technique which we have in dealing with propositions. And by "propositions" we mean such things as "It rains", "There are three chairs in the room", etc. English sentences about physical objects, sense datum propositions: this forms the nucleus of what we call propositions; and it is the practice or technique of using these expressions which is shown by the laws of logic.

If we said, "The law of contradiction doesn't hold" or "'$\sim (p. \sim p)$' is no longer true", we would be saying that we are not using "not" and "and" as negation and conjunction any more, or that "*p*" is not a proposition, or something of the sort.—Thus we might say that the laws of logic show what we *do* with propositions, as opposed to expressing opinions or convictions.

They are not unique in this. There are propositions regarded as synthetic *a priori*, like "A patch cannot be at the same time both red and green." [1] This is not reckoned a proposition of logic. But the impossibility which it expresses is not a matter of experience—it is not a matter of what we have observed.

We could give "This patch is red and green" a meaning; and you could even choose, among various meanings, the most natural one.—If I said, "This patch is red and yellow at the same time", this might at once suggest that it's *orange.* But a person who says that a patch cannot be red and yellow at the same time immediately has an objection to that. He will say, "This isn't what I meant." Asked what he meant, he will give an example, "It can't be red and yellow in the sense in which it can be red and soft." Or he might point at a pure red patch and a pure yellow and say, "It can't really have both these qualities fully and at once--as something does when it is red and oblong."

Suppose he says, "It *can't* be both red and yellow in the same way it *can* be both red and oblong." What does "in the same way" mean here? We might ask: What *would* you call 'being red and yellow in the same way it can be both red and oblong'? Does it necessarily contradict our previous interpretation of "being red and yellow" as being orange?—You want to continue the use of a picture, the picture "red and oblong"—but who says how to apply that picture in this case?

Is there a patch here that is both red and yellow?—What do you mean by "both red and yellow"?—"Well, you know—like both red and oblong, or both red and large." But which *is* the sense in which it is both red and oblong? Who says? Why shouldn't it be this—namely orange?

Suppose a man takes two points and draws a line through them:

1. See also *Zettel*, §§354–368.

He then shows me two points and tells me "Do the same"; and I draw this line:

Am I necessarily wrong? If he says, "No, I told you to do the same", I may say, "Well, this *is* the same."

Or suppose we take a circle and inscribe in it a pentagon, and then a square, and then a triangle. And we now say, "Go on and inscribe a biangle." He might perhaps draw a diameter. "Now go still *further:* inscribe a monangle."

He might draw some figure:

If we said, "But that is different", he might reply, "Well yes, of course. But then the pentagon was different from the square, and the square was different from the triangle."—What is the continuation of that line, and why shouldn't we say that this is?

Or he might say, "There is no such thing"—which would come to "I am not inclined to call anything the continuation of that line."

Similarly we are not inclined to continue the series "red and soft", "red and oblong", and to say "red and blue". Or we might continue it in that direction and say that a thing is red and blue if it is purple, but yet be disinclined to continue it to "red and green".

People say, "There is no such thing as reddish green." There is no reason why we shouldn't call *black* reddish green. Someone might object that we don't recognize in black the constituents of red and green. And there is something in this, of course.

It may be said that we recognize orange as reddish yellow

because orange paint comes from red and yellow paint. But mixing paints cannot in a sense show us that orange is reddish yellow. Why shouldn't there be a chemical reaction?

You might say, "That is not what we mean by mixing. We mean you use a colour mixer top." But suppose that when you spun it with red and yellow discs, the velocity made it go black. Would you then be inclined to say that black is a blend of red and yellow?

So we do not use experience as our criterion for orange being a blend of red and yellow. For even if the paints and the top gave black, we should not call black a mixture of red and yellow.

Turing: In this case, isn't one using "mixture" rather as one uses "multiply"?

Wittgenstein: Exactly so. That is just what I am driving at. We are *calculating* with these colour terms. The relation between: (1) mixing paints actually, or putting coloured discs on the colour mixer top, and (2) saying "red and yellow gives orange"—is the same as the relation between: (a) "two apples and two apples normally result in four apples", and (b) "$2 + 2 = 4$".

Suppose I showed you colours of all sorts of different shades, and then I asked you, "Which would you call *simple* colours?" Almost everyone will reply in the same way, and will pick out red, blue, green, and yellow, and perhaps black and white. We cannot say *why*. And if any of us were asked what simple colours these others were composed of, we'd all agree.

You might ask: Why do you call pure green a simple colour? or pure red? In fact, why do you call it *pure*? Nobody talks of "pure cream" normally. But "pure red" is understood. This is connected with the fact that probably if I taught you names for these 'dirty' shades here—if I asked you to paint my house the dirty yellow or dirty green of these pieces of furniture—you'd find it very difficult to remember these exact shades; whereas pure red or pure green you very easily remember. This is one reason why we use the expressions "pure red" or "pure yellow", and why we should all call the colour of this book a mixture.—Anyone would say, "This is composed of pure blue and white; it has pure blue in it, and white in it." Which is queer, because one doesn't see either pure blue or white in it.—Nor do we take it absolutely for granted that this colour can be obtained by mixing blue and white paints.

This has all been in order to get something clear about "laws of thought".

We might call a proposition like "There is no greenish red" a law of thought. Or "A thing can't be both green and red at the same time", or "A thing can't be both dark red and light red at the same time."

What would go wrong if we denied these laws? Nothing: except that it would upset our system. And that means simply upsetting *us*. For it doesn't mean that there is no longer any system.—Similarly, calling this: ⬯ a monangle would upset our system. We decline to continue it in this way. And this has all sorts of strong reasons. It hangs together very closely with the use we make of these things.

What upsets us, what we call a big or small change, an essential or inessential feature, hangs together with what we do with it—whether we need a new instrument to draw it with, and so on.—I don't mean to give an explanation.

The case against calling black reddish green is just this. We could imagine someone calling it that, but there are all sorts of reasons to be given against such a practice. It would come to building a system which would be decidely impractical.

How do we get convinced of a law of thought? Last term Turing said that a proof in mathematics may play either of two different roles, (1) of convincing someone; (2) of proving the truth: *really* to prove it, to make it indubitable—he used some phrase like that.

Let's talk about that. If I can talk about it, it will be the most important thing I have talked about. I want to show that there is some sort of muddle here.

It looks as if a man who is convinced by the first kind of proof really ought not to be convinced—he would not be convinced if he were really exact.—Similarly, Frege said that his proofs might seem rather lengthy, but if one really analyzed the whole meaning of the propositions, one would find that every single step was necessary—meaning that if you go two steps at a time you don't really get there.[2]

This goes together with the question as to what goes wrong if we assume a different logic to hold.

There are cases where we actually give a proof simply to quieten the conscience of the person: so that he can say he's had a proof. For instance, I used to learn engineering formulae; and in one case—torsion—there were three different ways of calculating which up to a certain point gave closely agreeing results. In that case, one might just say that a formula agrees with the results obtained experimentally; but people like to have a formula proved, so that they can feel on firmer ground. Here you can say, "I'm giving you a proof to convince you of this." But in this case there is another proof to show you that the conviction you've gained is right—namely, experience, the *experiment.* One can give a person a proof of some formula in physics to make him feel comfortable; but this will be no good unless the formula agrees with experience.

On the other hand, if thinking according to another law of thought does not mean we conflict with any experience or that we get into difficulties, but just that we use language differently—then there can be different kinds of mathematics. And if I convince you that you should draw conclusions in a certain way, "convincing you" meaning simply that you do draw them that way—then there is nothing wrong in your adopting this technique. If I then gave you a proof, I would not be giving you a foundation of it. There isn't such a thing as being convinced *rashly* in this case, as there is with an experiential proposition.

Suppose you have certain principles of logic, and by means of them you can deduce a certain further law. What does it mean to say that this law is *based* on those principles? It is based on them, *if* we actually adopt it *because* it follows in this way. The chain of reasoning may be our reason for adopting it. But it may *not* be.—It may be that at that point we like to use our laws in

2. *Grundgesetze,* I, vii–viii.

a different way—to make something which is an exception (which looks like one, that is).

You can't say we reason wrongly if, for example, at some point we do not accept the conclusion. We might say just "This is our logic: in this case *not* to accept the conclusion."

We multiply according to the system of cardinal numbers. If we did not recognize one of our normal multiplications of high numbers as being the *right* multiplication but at that point substituted another—you could not say we had violated the rules of mathematics, but that we had given a new rule. We've produced a new bit of mathematics.—We don't in fact do this, because we have no earthly reason to make such an exception. It would upset us in every possible way.

Intuitionism comes to saying that you can make a new rule at each point. It requires that we have an intuition at each step in calculation, at each application of a rule; for how can we tell how a rule which has been used for fourteen steps applies at the fifteenth?—And they go on to say that the series of cardinal numbers is known to us by a ground-intuition—that is, we know at each step what the operation of adding 1 will give. We might as well say that we need, not an intuition at each step, but a *decision*.—Actually there is neither. You don't make a decision: you simply do a certain thing. It is a question of a certain practice.

Intuitionism is all bosh—entirely. Unless it means an inspiration.

If you say: there is a proof which only convinces people of the truth, and on the other hand there is a proof which really makes the thing indubitable—there is something wrong. For to say that a proof convinces people might simply mean that they adopt that form of reasoning. It doesn't make it more indubitable. In fact the proof has to be adjusted in such a way as to get there.

"Proof that proceeds through indubitable steps": what is this? Isn't the indubitable step the convincing step?

All you've got for this step is a rule. And the use you make of this rule I suppose is the convincing use. There isn't a super-use.

You have a rule: for multiplying, drawing conclusions, etc. The rule is a certain expression, and then there is a certain technique of applying this rule. How is this technique given? Either in examples or not. Then what is the right application? Could it be the natural step and not the convincing step?

Suppose that I tell you to multiply 418 by 563. Do you *decide* how to apply the rule for multiplication? No; you just multiply. Probably no rule at all would come into your head. And if one did, no other rule for the application of the first rule would come into your head. It is not a decision; nor is it an intuition.

Malcolm: Couldn't we state the rules in a general way? For instance, we might say "3 X 3 is *always* 9."

Wittgenstein: We might say that, but in some higher multiplication give an exception. If you read the newspapers and see how people get round pacts, you should not be surprised at this.

We might explain it in several ways. We might, for instance, say, "Always, except in this one case." Or "Yes, 3 X 3 is always 9, but in this case write it as 7." Or we might find the use of "always" unnatural and give it up.—But the new multiplication is a new rule.

Russell says, roughly: "After all, it is not self-evidence which must guide one in the choice of primitive propositions. On the contrary, one is guided sometimes by the results which a given choice produces. Many primitive propositions are shown to be true by what follows from them." [3]—You may choose them because you want to get to a certain point. Not because they are indubitable.

XXV

The idea that there are two kinds of proof: 'the *real* proof'—the proof which gives a firm ground to the proposition, so that it is

3. Cf. *Principia Mathematica* (Cambridge, 1910), I, 13.

unshakeable and won't fall—and the proof that is to convince you. It doesn't make the proposition unshakeable—it only makes you believe that it is unshakeable.

This idea comes from a false view of what a proof actually does—and a false idea of the role which mathematical and logical propositions play.

Consider Professor Hardy's article ("Mathematical Proof") and his remark that "to mathematical propositions there corresponds—in some sense, however sophisticated—a reality".[1] (The fact that he said it does not matter; what is important is that it is a thing which lots of people would like to say.)

Taken literally, this seems to mean nothing at all—*what* reality? I don't know what this means.—But it is obvious what Hardy compares mathematical propositions with: namely physics.

Suppose we said first, "Mathematical propositions can be true or false." The only clear thing about this would be that we affirm some mathematical propositions and deny others. If we then translate the words "It is true . . ." by "A reality corresponds to . . ."—then to say a reality corresponds to them would say only that we affirm some mathematical propositions and deny others. We also affirm and deny propositions about physical objects.—But this is plainly not Hardy's point. If this is all that is meant by saying that a reality corresponds to mathematical propositions, it would come to saying nothing at all, a mere truism: if we leave out the question of *how* it corresponds, or in what sense it corresponds.

We have here a thing which constantly happens. The words in our language have all sorts of uses; some very ordinary uses which come into one's mind immediately, and then again they have uses which are more and more remote. For instance, if I say the word 'picture', you would think first and foremost of something drawn or painted and, say, hung up on the wall. You would not think of Mercator's projection of the globe; still less of the sense in which a man's handwriting is a picture of his character. A word has one or more nuclei of uses which come into every-

1. Page 4. "[Mathematical theorems] are, in one sense or another, however elusive and sophisticated that sense may be, theorems concerning reality . . ." Hardy does not speak of a correspondence to reality.

body's mind first; so that if one says so-and-so is also a picture—a map, or *Darstellung* in mathematics—in this lies a comparison: as it were, "Look at this as a continuation of that."

So if you forget where the expression "a reality corresponds to" is really at home—

What is "reality"? We think of "reality" as something we can *point* to. It is *this, that.*

Professor Hardy is comparing mathematical propositions to propositions of physics. This comparison is extremely misleading.

"To mathematical propositions there corresponds a reality" —if you take this in the sense of "Some mathematical propositions we affirm", then it is harmless but meaningless.

Or to say this may mean: these propositions are *responsible* to a reality. That is, you can't say just anything in mathematics, because there's the reality. This comes from saying that propositions of physics are responsible to that apparatus—you can't say any damned thing.

It is almost like saying, "Mathematical propositions don't correspond to *moods;* you can't say one thing now and one thing then." Or again it's something like saying, "Please don't think of mathematics as something vague which goes on in the mind." Because that has been said. Someone may say that logic is a part of psychology: logic treats of laws of thought and psychology deals with thought. You could get to the idea of logic as extremely vague, as psychology is so extremely vague. And if you oppose this you are inclined to say "a reality corresponds".

If you were to point out what mathematics is responsible to, then you would get the reality to which mathematics, in a sense, does correspond.

Here we see two kinds of responsibility. One may be called "mathematical responsibility": the sense in which one proposition is responsible to another. Given certain principles and laws of deduction, you can say certain things and not others.—But it is a totally different thing if we ask, "And now what's *all* this responsible to?" The axioms and the way of drawing conclusions may be said to be responsible to something, or not to be arbitrary.

When we speak of the responsibility of one proposition to axioms and laws of transformation, I have constantly stressed that given a set of axioms and rules, we could imagine different ways of using them. You might say, "So, Wittgenstein, you seem to say there *is* no such thing as this proposition necessarily following from that."—Should we say: Because we point out that whatever rules and axioms you give, you can still apply them in ever so many ways—that this in some way undermines mathematical necessity?

Von Wright: We oughtn't to say that; for the kind of thing we get in mathematics is what we call necessity.

Wittgenstein: Yes, one answer is: "But this is what we *call* necessity. We say '25 X 25 = 625 follows necessarily from so-and-so.' "

Or if a person says that this undermines mathematical necessity, you might ask, "What is your paradigm of necessity?—Show me first of all what you call necessity, and then we'll talk about whether this is necessity."

Now we have various paradigms in this case. One is *regularity*. Another: we say, "It's necessary that he'll come here"—we cannot get on without him; or something nasty will happen if he doesn't. So here if we say a thing is necessary, there must be something that goes wrong if it doesn't happen.—Or we might have a game in which some moves are necessary and not others.

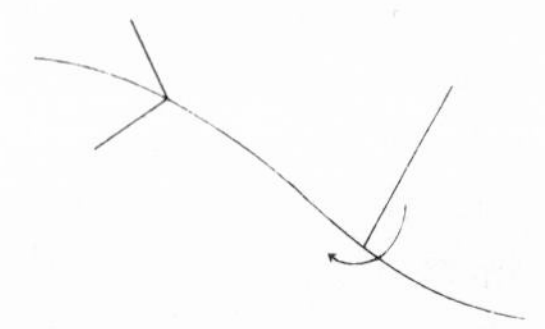

"*Here* the rules say you must turn right; *here* you may go whichever way you like."

What is necessary is determined by the rules.—We might then ask, "Was it necessary or arbitrary to give these rules?" And here we might say that a rule was arbitrary if we made it just for fun and necessary if having this particular rule were a matter of life and death.

We must distinguish between a necessity in the system and a necessity of the whole system. This is the point of von Wright's remark just now, that this is what we *call* necessary. He might have said that in this case it is not a question of whether the system as a whole is necessary.

We have to distinguish between different senses of 'necessary'. If we teach a calculus—and we have to multiply 21 X 14—we say the answer necessarily follows from certain axioms or premises. The question to ask is: Necessarily as opposed to what? Presumably, as opposed to the case where in our practice we leave it open what follows—or else it is a pleonasm.

This is analogous to an ethical discussion of free will. We have an idea of compulsion. If a policeman grabs me and shoves me through the door, we say I am compelled. But if I walk up and down here, we say I move freely. But it is objected: "If you knew all the laws of nature, and could observe all the particles etc., you would no longer say you were moving freely; you would see that a man just cannot do anything else."—But in the first place, this is not how we use the expression "he can't do anything else". Although it is *conceivable* that if we had a mechanism which would show all this, we would change our terminology—and say, "He's as much compelled as if a policeman shoved him." We'd give up this distinction then; and if we did, I would be very sorry.

To say "If you multiply these two, you necessarily get such-and-such a number", if it means anything *at all,* must be opposed to a case where there is no necessity. Or else it's a pleonasm to say you *necessarily* get this—why not simply say that you *get* it? —We might speak of getting something but not necessarily, in the case of a calculus in which you could get more than one answer.

With regard to "responsibility to reality": On the one hand you might say, "This conclusion is responsible to certain axioms and certain rules." This responsibility is based on our peculiar practice of using these rules. But then there is another question: as to whether such a system as a whole is responsible to anything. And to investigate this I tried to point out what *does* go wrong if we draw conclusions in a different way. We saw two things.

(1) We are then no longer inclined to use words as we do use them: for example, to use a certain word as negation—"there's no such thing"—and a certain word as a conjunction—"there's sugar and coal there".

If we give a word one particular partial use, then we are inclined to go on using it in one particular way and not in another. "Not" could be explained by saying such things as "There's not a penny here" or saying to a child "Must *not* have sugar" (preventing him). We haven't said everything but we have laid down *part* of the practice. Once this is done, we are inclined when we go on to adopt one step and not another—for example, double negation being equivalent to affirmation.

If the logical laws do not hold—we don't get the game we want to get, we don't play the game we want to play.

Suppose I said, "If you give different logical laws, you are giving the words the wrong meaning." This sounds absurd. What is the wrong meaning? Can a meaning be wrong? There's only one thing that can be wrong with the meaning of a word, and that is that it is unnatural. To give "not" the meaning of "and" and vice versa is not at all unnatural. But there are other things which are unnatural. For instance, we said we don't want to say "reddish-green". It is unnatural—unnatural for *us*—to use "red" and "green" in the way we're accustomed, and then to go on to talk of "reddish-green". And it is unnatural for us, though not for everyone in the world, to count: "one, two, three, four, five, many". We just don't go on in that way.

(2) If we allow contradictions in such a way that we accept that *anything* follows, then we would no longer get a calculus, or we'd get a useless thing resembling a calculus.

Suppose you had to say to what reality this—"There is no reddish-green"—is responsible. Where is the reality corresponding to the proposition "There is no reddish-green"? (This is entirely parallel to Hardy's "reality".)—It makes it look the same as "In this room there is nothing yellowish-green." This is of practically the same appearance—but its use is as different as hell.

If we say there's a reality corresponding to "There is no reddish-green", this immediately suggests the kind of reality corresponding to the other proposition. Which reality would you say corresponds to that? We have in mind that it must be a reality roughly of the sort: the absence of anything which has this colour (though that is queer, because, in saying that, we are saying just the same thing over again). It is superhuman not to think of the reality as being something similar in the case of "There is no reddish-green".

Now there *is* a reality corresponding to this, but it is of an *entirely* different sort. One reality is that if I had arranged for myself to call something reddish-green, other people would not know what to say. (Although I might appeal to examples in support of my use: to certain holly leaves which are red at one point and green at another and at a point in between they are a sort of irridescent black. I've often thought that if I had to call something reddish-green it would be that.)

"This is a flimsy thing to consider—whether one is *inclined* to say this or that." But it is no more flimsy than whether one is inclined to compare it with one thing or another thing, or whether one is inclined to use this picture or that picture.

I once knew a boy who talked of the 'dark notes' on the piano, not meaning the black notes but the low notes, although he had never heard them referred to as dark.[2] We might say, "He felt a similarity between darkness and low notes." If someone asked, "What *is* this similarity he felt? Where does it lie?", what could you say?—This *is* the similarity: that he wanted to say "dark". Isn't this what we call "noticing a similarity"? If we say, "He is inclined to use the word 'dark' ", this is like "He is inclined to use this picture ■ instead of that □."

This [inclination to call the low notes 'dark'] may be connected with all sorts of facts: that a child is frightened in the dark, not in the light; that he knows what a growl is, and is more inclined to be frightened by deep rumbling than by twittering.

What connexions we are inclined to make is (a) of the most enormous importance, (b) hangs together with all sorts of things.

If you were to say what reality corresponds to "There is no reddish-green"—I'd say: You may say a reality corresponds, only (1) it is of an entirely different kind from what you assume; (2) [what you have is a *rule,*] namely the [rule] that this expression can't be applied to anything. The correspondence is between this

2. Cf. *The Brown Book,* Part II, §4; *Eine Philosophische Betrachtung,* §§111, 123–124, in *Schriften* 5 (Frankfurt, 1970).

rule and such facts as that we do not normally make a black by mixing a red and a green; that if you mix red and green you get a colour which is 'dirty', and dirty colours are difficult to remember. All sorts of facts, psychological and otherwise.

We might ask, "Does any reality correspond to: 'A double negation gives an affirmation'?"

Think of "Two such-and-such things give such-and-such." Like: "If you turn this round twice . . ." "If you insert two pennies in the slot, so-and-so will happen." "Two turns of the handle produce such-and-such an effect."

We might think of two cases here. We might think of a light switch: turning it around once turns on the light, turning it around again turns it off. Or we might think of turning a match through 180°, then 180° again, so that the head faces the original direction.—Now there is a great difference between these two illustrations. The case of the match might be called a geometrical demonstration; the other might be called an experiment—to see how the switch works.—To put this in another way: the light might fail to work. But what would correspond to this in the case of the match? Nothing geometrical; what would correspond to it would be, say, the match breaking.

(Suppose someone says, "That space is three-dimensional is a matter of experience." What experiments would be made? Should we hold up three sticks at right angles and say, "Obviously we can't put another stick in at right angles to these"? What rot!)

Now suppose I turn the match around. What reality did I point to? What did I show you?

I might show you that the light switch goes on at every other turn, and not, say, at every third turn. But if you say I showed you that turning the match through 180° twice brings it back to the same position—isn't this just a matter of definition?

You could have a case of measuring. You might take a protractor, measure off 180°; you measure off 180° *again,* and turn it, and see whether it points in the same direction as before. This would be an experiment.—If it didn't point in the same direction, would you say the protractor was wrong, that it had expanded, etc.,—or would you say that in this case twice 180° does not bring you back to the same position?—So if you hold out the match and turn it round, if you say you are 'demonstrating something' —I don't know what you're demonstrating. You're turning a match.

"What reality corresponds to the proposition that if you turn a match twice through 180° it gets back to its original position?"

If this is a geometrical proposition, the reality which corresponds is: if we use a good protractor, then normally it brings us back, or more nearly back the better it is (where 'better' is determined by other criteria).

Yes—a reality corresponds to it, but it isn't of the kind you at first expect. At first you imagine that this is an experiment, like the experiment with the light switch. Then you discover that there is a reality corresponding to it but—if I may use the phrase—a much less clear reality: all sorts of things about protractors, etc., the fact that we can normally turn this round, and so on.

What you are saying is not an experiential proposition at all, though it sounds like one; it is a *rule.* That rule is made important and justified by reality—by a lot of most important things.

If you say, "Some reality corresponds to the mathematical proposition that $21 \times 14 = 294$", then I would say: Yes, reality, in the sense of experiential (empirical) reality *does* correspond to this. For example, the central reality that we have methods of representing this so that it can all be seen at a glance. In such a case as 21×14 nothing is easier than to lay out 21 rows of 14 matches and then count them; and then there is no doubt at all that *all of us* would get the same result. This is an experiential result; and it is immensely important. In such a case, if we looked at the things, we could easily notice if a thing vanished. We would all immediately agree that something had vanished, or that nothing had vanished.

I want to talk about the question whether one can justify the results of mathematical calculations by means of Russell's logic, or whether they depend upon certain quite different techniques; say the technique of being able to compare two numbers of objects in a certain way. For instance this: *Principia Mathematica* has been printed in a few thousand copies. We say they all contain the same proofs. There is a way in which these copies have been produced, and this has been checked; and this satisfies us.

XXVI

If one talks about a reality corresponding to mathematical propositions and examines what that might mean, one can distinguish two very different things:

(1) If we talk of an experiential proposition, we may say a reality corresponds to it, if it is true and we can assert it.

(2) We may say that a reality corresponds to a *word,* say the word "rain"—but then we mean something quite different. This word is used in "it rains", which may be true or false; and also in "it doesn't rain". And in this latter case if we say "some phenomenon corresponds to it", this is queer. But you might still say something corresponds to it; only then you have to distinguish the sense of "corresponds".

If you say, "Something corresponds to the word 'red', namely this colour"—how does it correspond if you say (truly), "There's nothing red in this room"? And you might also have "There's nothing red in the world."

Von Wright: It doesn't seem to me there is a very big difference between correspondence in the case of a sentence and correspondence in the case of a word.

Wittgenstein: There is an enormous difference.—Suppose I spoke of the reality corresponding to this sentence. I may mean two entirely different things: (a) I might mean that the sentence is true; (b) I might mean that there is a reality corresponding to the words in it—that is, that the sentence has a meaning. And these two things are entirely different. In the one case, by saying "A reality corresponds to so-and-so" we are affirming a sentence.

In the other case not; we don't say anything about anything.

If I say, "A reality corresponds to 'rain' "—what sort is this?

Suppose I point to a chair and say "This is green." I might be said to point to the reality which corresponds to "green". Showing the reality which corresponds to a word, is giving the word a meaning. (Now if you say, "Everything green has vanished from the earth"—) This correspondence may be called *a correspondence of grammar.*

If you say, "This reality corresponds to this word"—this is a sentence of grammar; you are giving a grammatical explanation. Whereas if you say, "A reality corresponds to 'There are six people in this room' "—this is not a sentence of grammar at all; you are affirming a proposition. This is an essential difference.[1]

Suppose we said, "A reality corresponds to the word 'two'."—Should we say this or not? It might mean almost anything.

"A reality corresponds to the word 'perhaps'."—Does one, or not? You *might* say so; but nobody would.—Or to "or", or to "and". It is unclear what reality we should say corresponds here. I don't mean we do not give them a meaning. And I *might* do something like this: I say "It's very uncertain whether Smith will come"; I draw a picture of his entering; then I point and say "This is 'perhaps'."

Similarly with "A reality corresponds to 'two'." I might point to something. [*Wittgenstein raised two fingers and pointed to them.*] But you wouldn't know what the reality is which corresponds; this isn't clear.

The point is this. We *can* explain the *use* of the words "two", "three", and so on. But if we were asked to explain what the reality is which corresponds to "two", we should not know what to say.—This? [*He indicated the two fingers.*] But isn't it also six, or four?

We have certain words such that if we were asked, "What is the reality which corresponds?", we should all point to the same thing—for example, "sofa", "green", etc. But "perhaps", "and", "or", "two", "plus" . . . are quite different.

1. (From "Suppose I".) The three versions of these two paragraphs are quite different. The text includes material from all three.

If a man asks, "Does no reality correspond to them?" what should we say? How should we explain this feeling that there is a reality corresponding to these words, too?—He means "Surely we have some use for them." And that is obviously true.

Suppose I say, "We have some use for negation. Why?" Could one answer, "Because there are false propositions"? Well, we ought to answer that it's an ethnological fact—it's something to do with the way we live. We bar certain things; we don't let a man in; we exclude certain things; give orders and withdraw them, make exceptions, etc.

So with these words "and", "or", etc., we can say that the reality which corresponds to them is that we have a use for them.

What I want to say is this. If one talks of the reality corresponding to a proposition of mathematics or of logic, it is like speaking of a reality corresponding to these *words*—"two" or "perhaps" —more than it is like talking of a reality corresponding to the *sentence* "It rains". Because the structure of a "true" mathematical proposition or a "true" logical proposition is entirely defined in language: it doesn't depend on any external fact at all.

I don't say: "No reality corresponds."

To say "A reality corresponds to '2 + 2 = 4' " is like saying "A reality corresponds to 'two'." It is like saying a reality corresponds to a rule, which would come to saying: "It is a useful rule, *most* useful—we couldn't do without it for a thousand reasons, not just *one.*"

You might say: Mathematical and logical propositions are still *preparations* for a use of language—almost as definitions are. It's all a put-up job. It can all be done on a blackboard. We just look at the signs and go on here; we never go outside the blackboard.—The correspondence of mathematical propositions to reality is like the correspondence of negation to reality. It is all entirely *independent* of the other correspondence with reality, the correspondence of "it rains". It's like the correspondence of a word to something used in an ostensive definition.

In mathematics the signs do not yet have a meaning; they are *given* a meaning. "300" is given its meaning by the calculus—that meaning which it has in the sentence "There are 300 men in this college." In the sense in which we might say "This is a chair" gives a meaning to "chair". Similarly in logic, the signs don't yet *have* a meaning, but are *given* a meaning: the meaning of "not" is given in "$\sim\sim p = p$".

If we say "300" has a meaning in "There are 300 men in this college", then in mathematics it does not *have* this meaning, but is given it.

If I wanted to show the reality corresponding to "$30 \times 30 = 900$"—I'd have to show all the connexions in which this transformation occurs.—Notice the difference between asking, "Is there a reality corresponding to '$30 \times 30 = 900$'?" taken alone, and saying this of it as a proposition in a system. Taken by itself we shouldn't know what to do with it: it's useless. But there is all kind of use for it as a part of a calculus. If we had a different calculus, "$30 \times 30 = 900$" might not have any meaning.

You might say: Mathematics and logic are part of the *apparatus* of language, not part of the application of language. It is the whole system of arithmetic which makes it possible for us to use "900" as we do in ordinary life. It *prepares* "900" for the work it has to do.

In this sense, mathematical propositions do not treat of numbers. Whereas a proposition like "There are three windows in this room" *does* treat of the number 3.

Putting this in a different way: Suppose I say "Prince has blue trousers"; that is a proposition about the trousers. We could extend the use of "about" to colours. What are we to say are propositions 'about blue'? We could say two quite different things. We might say "There are blue books here" is a proposition about blue; or, thinking that this is not about blue because "blue" is only an adjective there, we might say "Blue is darker than yellow" is a proposition about blue.

I say the way I'll go is the first way: "Prince has blue trousers" is an example of a proposition about blue, and "Blue is more similar to purple than to yellow" is not. The latter type—like "A sofa is longer than a chair"—is grammatical. And here there is great danger.

Take the words "one metre". What is a proposition about one metre? "This chair is one metre high." But "A metre is about 39 inches" is not about a metre.

What about "two"?

"2 + 2 = 4"—but this *isn't* about 2: it is grammatical.

Turing: Isn't it merely a question of how one extends the use of the word "about"?

Wittgenstein: That is a most important mistake.—Of course you can say mathematical propositions are about numbers. But if you do, you are almost sure to be in a muddle. Because you don't see that what is about 2 in the sense in which a proposition is about a sofa, is never a mathematical proposition.

If someone asked, "Which propositions of Euclid are about triangles?", I have no objection at all to saying that the propositions on p. 30 are about triangles, those on p. 40 are about circles, etc.

I don't say it's *wrong* to say that mathematical propositions are about numbers, that the other way of speaking is right. I only want to *point it out.* Because unless you see that there are the two ways—you are likely to be misled.

I have pointed out a source of confusion. I do not mean there is a constant confusion there. If you say, "The proposition so-and-so in Euclid is about a circle", there is no confusion whatever. But as soon as you talk about the reality corresponding to mathematics, there is an enormous confusion if you do not see that "being about" means two entirely different things.

This brings an entirely different sense of how a reality corresponds to mathematics. Because now, if "30 × 30 = 900" is not a proposition 'about 30', you will look for the reality corresponding to it in an entirely different place; not in mathematics but in its application. (Contrast what you'd do with "I have 30 handkerchiefs".)

If you have a mathematical proposition about $\aleph_0$, and you imagine you are talking about a realm of numbers,—I would reply that you aren't as yet talking about a realm of anything, in the most important sense of "about". You are only giving rules for the use of "$\aleph_0$".

You are developing the mathematics of it. And you have now to ask: in which *non*-mathematical propositions is it used? If you want to know the realm to which it points, you have to see in what sentences we use it.

As soon as you do this, you get an entirely different picture of what you have been doing. At first, we picture ourselves flying to the end of the cardinal number series and beyond; this comes from thinking of mathematical propositions as the *application* of numbers. We get an entirely different picture if we consider it the other way: the statement that John has mastered $\aleph_0$ multiplications will mean he has mastered a certain technique of multiplying. And now we see we haven't been flying anywhere.

If you want to understand a word, we always say: "You have to know its use." It is immensely important that to the great majority of words there correspond certain pictures which in some sense or other *represent* for us the meaning of the word.—In one sense this is clear: a picture of a chair may stand for the word "chair" and so on. In the case of "chair" that picture is of enormous use; it is actually used to explain the word—or to explain what a 'Chippendale' is, for instance. Once we are shown this, we are sure to use the word in the same way.

But in other cases these pictures are more or less misleading or useless. For instance, the picture of a 'particle' can be extremely misleading—where the expression is no longer applied in such a way that this picture is of any use. You cannot explain how "particle" is used in physics by pointing to, say, a grain of sand; indeed, if you did that, you'd make a hash of it.

There are many such pictures in mathematics.—A calculus for us is something we have learned. Every one of us, whether he is a mathematician or not, first learned to multiply in this way—as we do when we get $30 \times 30 = 900$. So that is what is first and foremost in everyone's mind, directly he hears of multiplication. And similarly with addition. So now if someone says $\aleph_0 + 1 = \aleph_0$, we get the old picture of adding something to something. Whether this picture is misleading or not will depend on the further use one makes of, say, $\aleph_0$.

If I say "the cardinal number of all cardinal numbers" or "the cardinal number of the concept 'cardinal number' "—this is an expression which reminds us of *ten* thousand other expressions, like "the cardinal number of all the chairs in this room". It conjures up all sorts of pictures—for example, the picture of an *enormous colossal* number—which gives it a great charm. And to say that there is a subject treating of this number and of greater numbers—we are dazzled by the thought. (It is not only children that ask, "What's the greatest so-and-so?") Then if you realize that by forming the expression "cardinal number of the concept 'cardinal number' " you haven't yet given it any application at all, you see that you have as yet no right to have any image. Because the imagery is connected with a *different* calculus, $30 \times 30 = 900$.

We say our space is three-dimensional. Then someone says, "Now imagine something four-dimensional." You might take three coordinates and go around seeing that there is no right angle not already filled.—Suppose someone says, "We have an imagery of one dimension, and of two dimensions, and of three dimensions. Now go on to four dimensions." And the reaction might be "Good Lord: It's terrific!"—But I'd say: If you think it's terrific, if it astounds you or even has a charm for you, it is because you have the wrong imagery.

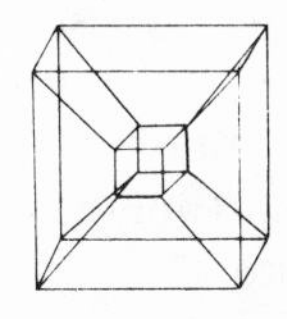

Suppose I say this is a four dimensional cube.[2] Then that's all right; but there is nothing terrific about it. It is pedestrian.

If you want the right image for $\aleph_0$, you mustn't form it from mathematics. If you say "How terrific!", if your head reels—you can be sure it is the wrong image. It's not terrific at all.

If we say of a child who has learned to multiply that he has learned $\aleph_0$ multiplications, then we have the right imagery. But

2. Cf. *Zettel,* §249

not if we have the image of a line of $\aleph_0$ lime trees, which we cannot see the end of.

This business about imagery comes from the fact that a mathematical proposition is not *about* its constituents in the sense in which "The sofa is in this room" is *about* the sofa.

I know what I'll say now will sound awful. Certain parts of mathematics tend to be regarded as specially deep. Professor Littlewood in one of his books talks like this: The part of mathematics with which he is dealing is extraordinary, not in that it contains complicated calculations, but that the depth lies in what is said. The beauty of the subject lies not in the calculations, which are as simple as anything, but in the *meaning* which they have.[3]

Now I say that the only meaning they have in mathematics is what the calculation gives them. And if it's simple, it's simple.

One might say, "Boys up to twenty learn complicated calculi, but you need an educated brain for this simple one, for these highly abstract notions." As though here we had to see through the calculations to a depth beyond.—This I want to say is most misleading. The calculus (system of calculations) is what it is. It has a use or it hasn't. But its use consists either in the mathematical use—(a) in the calculus which Littlewood gives, or (b) in other calculi to which it may be applied—or in a use outside mathematics. It is as pedestrian as any calculus, as pedestrian as the four dimensional cube. If you think you're seeing into unknown depths—that comes from a wrong imagery. The metaphor is only exciting as long as it is fishy.

Turing: What Professor Littlewood said was right, in that one needs a different technique in that branch of mathematics, in order to find one's way about.

Wittgenstein: Yes.—Where does a calculus (system of calculation, take its interest from? It may be (a) from an application of it, (b) from the pictures which go with it and which arise from certain analogies which this calculus has to other calculi.

3. Wittgenstein may have been referring to *The Elements of the Theory of Real Functions* (Cambridge, 1926), p. v.

Take the infinitesimal calculus. The idea that it deals with infinitely small things gave it a charm quite apart from its usefulness. And although this idea has now been abandoned, it still has a charm.

When people criticized the earlier idea, they sometimes said, "When we look into the calculus, we find that there is nothing infinitely small there." But what did they expect to find? Why were they disappointed? What does it mean to say that the calculus doesn't treat of anything infinitely small? or that "there isn't such a thing as anything infinitely small"? or: "We look at these calculations and we don't see anything infinitely small"?—This might be contradicted. Why shouldn't you say *dx* was infinitely small?

The point is: First, "infinitely small" has no clear image corresponding to it, and you *could* say [the calculus] does treat of something infinitely small. And secondly, instead of saying it doesn't treat of anything infinitely small, what one ought to say is that *it doesn't treat of small things at all.*

Similarly with "the infinite". "We aren't talking of anything you would call *big,* and therefore not of anything infinite."—But as long as you try to point out that we are not treating of anything infinite, this means nothing, because why not say that this *is* infinite? What is important is that it is nothing *big.*

When one is a child, "infinite" is explained as something huge. The difficulty is that the picture of its being huge adheres to it. But if you say that a child has learned to multiply, so that there is an infinite number of multiplications he can do—then you no longer have the image of something huge.

If one were to justify a finitist position in mathematics, one should say just that in mathematics "infinite" does not mean anything huge. To say "There's nothing infinite" is in a sense nonsensical and ridiculous. But it *does* make sense to say we are not talking of anything huge here.

A member of the class: Even when one says that a child has mastered an infinite technique, there is even there an element of hugeness and one has the idea of something huge.

Wittgenstein: Yes, but the idea of hugeness in that case comes only from the word "infinite" and not from what it's used for. By watching his work, we shouldn't get the idea of anything huge. The teacher does not say to himself, "Ah, fancy these boys of ten and eleven having such vast knowledge!"

XXVII

I want to talk about the relation between logic—what Russell and Frege mean by logic—and arithmetic.

It may seem queer that Euclidean geometry talks of 'length' and 'equality of length' and yet not of any method of comparing lengths. Especially since "this length is equal to that" changes its meaning when the method of comparison is changed.

This isn't pointing out any shortcomings of Euclidean geometry. It only shows that in a most important sense Euclidean geometry does not talk of lengths. If you say, for instance, that a circle is the locus of all points equidistant from a given point—this might mean anything. A circle might look like this:

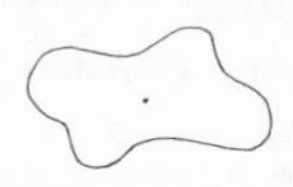

if one were given a suitable method for measuring equality of length. (The easiest way to imagine this is seeing a circle in a distorting mirror. Of course, even if a circle does look like that, it will still be true that a straight line only cuts it in two points; for a straight line will look correspondingly queer.)

But we could say that Euclidean geometry gives *rules* for the application of the words "length" and "equal length", etc. Not *all* the rules, because some of these depend on how the lengths are measured and compared.

Similarly, arithmetic doesn't talk of numbers, in that it doesn't give us any method of finding a number or of comparing the number of these with the number of those—but gives us rules for the use of number words.

There are many different ways of finding and comparing numbers. We can, for example, compare numbers by the eye.

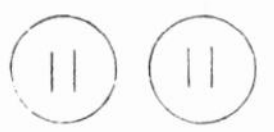

That there are the same number of dashes there and here, nobody gets by counting.—Then from a certain point onwards we count. Or we can one-one correlate the dashes.

A different way is, for example: if you know a poem by heart and you say it twice, you say you have said the same number of syllables—you didn't count them and you didn't one-one correlate them. Or again: if you have a complicated ornament and a duplicate of it, you say this is the same ornament as that, and it has the same number of angles, or intersection points of its lines. Or again: if you whistle "God Save the King", and I do, we say we've whistled the same number of notes.

Normally all these ways of counting agree. We might count money by whistling "God Save the King"—we have the same number of shillings if we reached the same point in it. And then if, for example, we *counted* them, the two numbers would agree. And if we put stacks of shillings side by side, they'd correspond one-one.

This is so. But it *needn't* be so.—What if we don't get the same result? We'd say, "I've made a mistake" or "A shilling has vanished", etc.

We might try to find the mistake. But now there are two possibilities: (a) that we find the mistake (you whistle the tune twice, and make a record, and find that the second time you left out a note); (b) that we don't discover the mistake.—But then is it not even a question whether we should call it a mistake?

Why are we so inclined to say that there is a mistake? (1) it hardly ever happens that there is a discrepancy; (2) when it does happen, we can nearly always find the mistake.

But it is not clear what we should accept as clearing up the mistake, why we should take this as reliable and the other not.

If the gramophone record shows that a note was missing, we trust the record and not our memories. But why say that the gramophone records were the reliable thing? Well, if I trusted my memory as opposed to this, I'd have now to assume all sorts of things—an extremely awkward physical theory. You can do it, but—

I said that arithmetic is not concerned with the way in which we arrive at numbers. This isn't a shortcoming. It wouldn't come in here except for a particular reason. It might be said that although the way we count apples doesn't come into arithmetic, the way we count something within *arithmetic* does come in.—But we will leave this for a moment.

That there is hardly ever any discrepancy between ways of counting, and when there is we are able to clear it up usually, is of immense importance. If these discrepancies happened more often, then all sorts of things would happen.

For instance, in certain circumstances, we might say that counting by "1, 2, 3, 4, . . ." is *unreliable.* Suppose it is important to count shillings and get them to stack up to the same height. We might say that counting is unreliable, but that correlating by strings is reliable. That would mean that we should give up counting for certain purposes.—Or suppose that when we invite people to dinner and we count the names and the butler counts the chairs, then invariably there is a mess because some are without chairs, or there are too many chairs; and no one can ever find out whether more guests have come or a new guest has turned up, . . . —"This would be an odd state of affairs." Yes. But suppose: "It would be an odd state of affairs if one couldn't say what the weather would be tomorrow." That isn't odd at all. And the other wouldn't have to be.

Arithmetic and logic. If one had to prove by Russell's logic that 4,000,000 + 3,000,000 = 7,000,000, one does it by saying roughly this:

If 4,000,000 entities—that is, (x, y, . . . up to 4,000,000)—satisfy the function ϕ, and only 4,000,000; and 3,000,000 satisfy ψ, and only 3,000,000; and no ϕ's are ψ's—then the sum of the two numbers is the number of things which are either ϕ or ψ.

(If I have three apples in this hand and four in that and none which is in both, then in: "either in this hand or in that hand" we have the sum of the two numbers.)

Now there is of course such a thing as a mistake in your calculation. If you want to do this calculation by Russell's logic, you have to see that there are 3,000,000 variables in the first bracket before the multiplication sign and 4,000,000 in the other, and 7,000,000 in the bracket after the implication sign:

$$(3 \cdot 10^6)\ (4 \cdot 10^6) \supset (3 \cdot 10^6 + 4 \cdot 10^6)$$

And the proof comes to *a cancelling out.*

Actually no one would prove it in this way—by cancelling out; and one is inclined to say that you might easily go wrong.

There are other ways of doing it—through definitions, etc. But then the question is: Must we get the same result by both methods? And then which result is correct? and why trust one method rather than another?

Suppose the scheme of implication were somehow represented by electric wires. If they are wired correctly, a phenomenon occurs—say a certain bell rings. In Russell, the point is that the whole thing should become a tautology. We might speak of 'weighing' the numbers on the balance of a tautology.

Say a galvanometer needle points to nought if the wiring is correct: if it is not, there is some deflection.—If you *counted* to see whether the wiring was correct, your opinion would be confirmed by what happens on the galvanometer. The result gives us an added check: we agree, and the needle did what we expected.

What I want to say is that there is no galvanometer needle here. The whole point of the simile is that it is a bad one. It looks as though one could take the tautology as a criterion as one does a needle. But it isn't so at all. It is vice versa.

"If this were all written out correctly, or if we counted it correctly, it would give a tautology." But this is not the galvanometer business.

Turing: Doesn't this come to saying that counting by making tautologies is more unreliable than counting in the ordinary way?

Wittgenstein: No, it does not come to that. These are two entirely different ways.

It is not a question of its being unreliable. For all I know it may be perfectly reliable. Whether we can say arithmetic rests on—or is—logic: that's the point. That it might possibly be done that way, is a different matter. But the point is that *we* regard a thing as a tautology by an entirely different method. We introduce new principles. And it is not enough to say that we can make our principles agree with Russell's principles.

The idea that there is a science, namely logic, on which mathematics *rests.* I want to say it in no way rests on logic. And the fact that you can make logical formulae agree with it, in no way shows that it rests on logic.

We might count in a different way, and thus get from Russell's logic quite a different arithmetic. It might happen that our ordinary calculations gave one result, and our step-by-step calculations in Russell's logic always gave another result. And then it is not clear what we'd say.

We have normal ways of finding whether the numbers on both sides of the implication sign are the same. And this does not depend at all on Russell's principles; on the contrary, they depend on it. If we didn't have such ways of comparing the different sides, we shouldn't know what to call a tautology.

There might be reasons for saying that certain multiplications people make are always slightly wrong. There are two biggish numbers; and whenever someone multiplies these, the correct result is always one greater than the result he gets. No matter how closely we look, we never get to the mistake. "Then how did they get the idea that there was a mistake?" There might be all sorts of reasons: they take planks corresponding to the numbers multiplied (so many rows of so many) and planks corresponding to the number of the result, and they always find a discrepancy. Then instead of saying that one has vanished, or that one has split into two, they say there was a mistake in the calculation and add one.

One fact makes this seem more fantastic than it is: When Russell calculates, he never brings in large numbers. And we see that $3 + 2 = 5$ in a different way—not by counting. So it is more difficult here to imagine the sort of mistake I'm talking about.

If we take "Arithmetic is based on logic", we might think this meant that *our* arithmetic and no other follows from Russell's logic. You might say, "This is still inexact. You get any other arithmetic also. Only if you give Russell's definitions, then you get ours only. If you have another arithmetic, you are working with other definitions of 'cardinal number', 'if—then', 'and', etc. You might get, say, $3 + 4 = 6$, but then the signs just have a different meaning. When we mean by 'cardinal number' (for example) what Russell says it means, this arithmetic must follow."

This is what I deny. I'd say that not even this arithmetic follows from Russell's calculus—no more than any other.—What Russell does is to give a certain calculus for "if", "and", "not", etc.; and as far as these expressions occur in mathematics, this will hold. If Russell's calculus is to be merely an *auxiliary* calculus, dealing with "if"s and "then"s etc.,—then it is all right. But that is not what it is meant to be.

Turing: Do you mean that when we do arithmetic we don't do all this stuff of Russell's?

Wittgenstein: No. We might do it or we might not do it. For all I know the Martians may teach their children *Principia* and then the children multiply. But they might keep *Principia* and say that 20×30 is not 600 but 601, and have generally a quite different arithmetic.

If I give you a calculation to do, you say that you will do it by *Principia.* But what if I do it in the ordinary way and get a different result? How do we decide which calculation is correct?

Turing: It is just like any of the other pairs of ways of counting.

Wittgenstein: Exactly; that is the whole point. The Russellian method is just one method, like any of these other methods.

We *base* our calculations on the agreement of all sorts of things. And we might trust one thing although the other disagreed with it—we might even have to say, for example, that Russell's logic gives wrong results.

We cannot say that arithmetic is based on Russell if Russell is based on arithmetic.

Is the truth of our arithmetical propositions based on logical truth? Or what is the relation between them?

Turing: Russell's definitions show us the *point* of having these ideas of addition and finite cardinals and so on.

Wittgenstein: Yes—and it is just that that I want to deny.

What do Russell's explanations do? What do Russell and Frege actually do?

Take Frege's statement that a numerical statement is a statement about a concept. This means that if we say, for example, "There are five nuts on the table" the *five* is not predicated of a heap but of a concept. We don't say that what we see here has the property *five;* because what we see here may have *any* number —one or a million; is it the number of atoms, for instance? But the *concept* "nuts on the table" has the property *five.*

This is a very great clarification as far as it goes.

Again, take: "I met a man" and "I met John". These look very much alike; it looks as though "a man" were like "John".—Frege expressed this in an entirely different way from Russell. Russell expresses "I met John" by "$f(a)$" and "I met a man" by "$(\exists x).f(x)$".—This is in a way similar to what Frege said about predicating the number, not of the heap of nuts, but of the concept; because this also cleared the grammar enormously and made certain misunderstandings impossible.

Take another case: Frege, when he had said that a numerical statement was about a concept, went on to say that a numerical statement was predicating something of a predicate. He would say that "a man on this sofa", for instance, is a predicate; because we say "Turing is a man on this sofa", just as we say "This sofa is green". And so when we say "There are two men on this sofa", we are saying that the predicate "man on this sofa" has the predicate "two".—This is both a clarification and a confusion.

There is a temptation to talk of a predicate of a predicate. One had the word "predicate"; and at first no one would have talked of a predicate of a predicate. People used to talk of subject and predicate in logic. And then Frege said, "We have a predicate of a predicate." One of the great things was the jingle. It was a grand discovery. And if you like it, you had better stick to it. (Compare "A class of classes".)

Turing said Russell's definitions make clear the *point* of the use of these words. But they do so only up to a point. Partly they confuse things.

It was a clarification to give a symbolism distinguishing between "I met a man" and "I met John". But there was a confusion too.

Take the case of 'predicate'. With the idea of a predicate, goes the idea of a *property.* For a property is almost the simplest form of predicate. Suppose I say, "This sofa is green", then the predicate is "is green". If I then ask what it is that *has* the property 'green', you would imagine something like a colourless sofa.

Suppose I say, "Turing is six feet tall." Then it is clear what the subject is and what the identity of the subject is. If I don't know how tall he is, I might ask him to stand up and then measure him. And we know what it would be like to say, "Now he is six feet tall; ten years ago he was five feet tall." The identity of the subject is given by the path he has described since he was a boy.

Similarly if I say, "This chair is four feet high": I know when I would call it this chair. For instance, if there is another chair just like this one, I distinguish them by the paths they follow. If it were possible to make two chairs coincide and separate again, as one can with shadows, then it would be unclear which is 'this chair' and which is 'that chair': what the criterion would be.—In fact, I might say, "This chair has so many studs"; and it would be clear what I would mean by "It has so many now, and it had so many (a different number) before."

Russell says "I met a man" means "There is an *x*, such that *x* is a man and I met *x*." Now what is the *x* here which is a man?

"*x* is a man" may have any number of different meanings. For instance, if I see a face, I might say, "This isn't a woman, this is a man." Or in the fog: "This isn't a lamp post, it's a man."

But we use this predicate business quite differently from the way Russell supposes that we use it.

Suppose I say, " 'Circle' is a predicate because we can say so-and-so is a circle." We don't know from this what we predicate circularity of. It may be of a thousand different things; and the sentence may have a thousand different meanings.

If I draw: ○ and say, "This is a circle"—what is it that is a circle? You could give it a hundred different interpretations.—What *is* a circle must be something which might not be a circle. If I say that *this* is no longer a circle, what would that mean? What's 'this', then?—In certain circumstances we might say, "It's contracted into an ellipse." But if it vanished and an ellipse appeared, we might not know whether to call it the same thing.

Suppose we say, "There is a circle in this square." Russell says this means "There is an *x* such that *x* is in this square and *x* is a circle." Now what in the hell is this *x*?—And what if we say "Everything in this square is a circle"? What the hell is to happen to this? "For all *x*, if *x* is in this square *x* is a circle." What does this mean?

All this symbolism comes from ordinary language. It could have been written in English or German, except for a few dodges like brackets and dots.—It's all right as far as it goes. But apart from that, it doesn't clear anything up. In fact, it makes confusions. I do not mean it is valueless. But it does not show the *point* of anything; it leaves everything as it is. It makes language a *trifle* more explicit, leaving all the confusions. It makes *certain points* clearer. It does not go into detail. It avoids *certain* limited confusions. And I don't think that doing this work was a simple thing; quite the opposite.

It translates arithmetic into a language in which we see certain points which we did not see before, and get into certain confusions which we would not have got into before.

When Turing said that Russell's definitions make clear the *point* of arithmetic, he means: Russell's explanation makes clear, for instance, the connexion between the addition of two numbers and the disjunction of two concepts. "2 + 3 = 5" doesn't mean that you put 2 here and 3 there . . . , but that if a concept has 2 and another concept has 3, the concept which is the disjunction of the two has 5. It makes clear in a way what it means to say "in both rooms together"—that is, in either this room or that room. So far so good. It shows a relation between addition and "or". This clears matters as far as it goes. But what in a particular case we are to call the sum of two numbers is not in any way clear. There are still all sorts of possibilities of interpretation. We have simply made a connexion between '+' and 'or'—which may be very important.

What Russell and Frege do is to make connexions between English and German words "all", "or", "and", etc., and numerical statements. This clears up a few points. But that we should actually then say, "3,000,000 + 4,000,000 = 7,000,000", does not follow from this. We could go on as we like—while we kept the same use of "all", "and", etc.—We might say, "Russell's principles are all right, but we can't calculate with them." And this would not mean that we ought to have different definitions.

Turing: Doesn't it come to saying this: if I have an ordinary method of counting, and then find another method which in fact gives the same results, I cannot really stick to the first method, because I'll be inclined to check it by the second—just as one might use the ordinary method of multiplication to check Russell's method.

Wittgenstein: Yes.—If we suppose we have two ways of multiplying, Russell's and ordinary arithmetic, is there anything which gives one a preference over the other? It's just that now we calculate the truth and falsehood of propositions of logic in a new way—in a non-Russellian way. We use each to check on the other. We might say that Russell's method is perfectly all right, but neither is more *fundamental.*

We might say: We *interpret* one in terms of the other—which means that we have a way of translating from one to the other. But this works both ways. If you say that Russell gives the point of arithmetic, you can also give the point of Russell by translating Russell into arithmetic. You could give a new mathematics a point by translating it into Russell—that is, into the language of "or" and "not". And further, you might never be able to translate arithmetic into Russell's language, and yet you might nevertheless be satisfied that you could translate it.[1]

It is a very queer thing to say that one can check up on logic by some other way. If you regard logic in a particular way, it seems the queerest question on earth to ask: "How can you check up on logic?" For what is meant by it? You ought not to be able to check up on logic, ought you? For one always thought that logic was the foundation of everything else. And if it really is the foundation, then it must be by logic that you check everything else.

This is the same problem as how we can have a proof—say by mathematical induction—a proof by means of a short cut. For instance, we can prove $f(1000)$ by proving $f(1)$, and that if $f(1)$ then $f(2)$, and that if $f(2)$ then $f(3)$ and so on. Or else, having proved $f(1)$ and $f(n) \supset f(n+1)$, one can make a short cut. And this is the queerest thing in the world: that one should have a short cut through logic. For if the proof of the proposition is the step-by-step proof, how can anything else be a proof of it? How can it be certain that the one will reach the same result as the other? Aren't we really rash?

How can there be a short cut through logic? A proof ought to be a proof, and everything cutting it short should be rash.

This is most important. It's puzzled me more that I can say. For the moment I will leave you puzzled.[2]

1. (From Turing's remark) The three versions of this passage overlap very little. The version given leans heavily on M for its structure, but also draws on the other two.

2. Wittgenstein returned to the subject of mathematical induction in Lecture XXXI.

XXVIII

One thing that was said was that Russell had given an interpretation for arithmetic, and that he gave a definition of cardinal number from which the cardinal number calculus follows. The difference in behaviour between the cardinal number 3 and the rational number 3 follows from the difference in their definitions.—I tried to say that in a sense Russell had given an interpretation by tying up primitive ideas of arithmetic with primitive ideas of logic—thus tying up arithmetic and propositions.

Whenever we use arithmetic we use it in connexion with *sentences.* It seemed as though Russell had supplied a calculus which was immediately applicable. It rests on propositions: propositions being absolutely fundamental and the last thing next to reality. Every child knows what a proposition is.

When logic has been conceived to be a calculus—as for example by Leibniz—it has seemed to be *the* calculus par excellence, absolutely indubitable and depending immediately on reality. (Dedekind's view: that arithmetic is an extended logic; but Dedekind did not show the connexion between logic and arithmetic as Frege did.)

There were two ideas in the Russell-Frege development:

(a) They saw the great similarity between the role which arithmetic plays and the role which logic plays. We make use of each (of logic and of arithmetic) in making a transition from one *material* proposition to another. We can say that arithmetical propositions are laws of thought in the same sense in which logical propositions are. So far this was correct.

(b) It seemed as though the laws of logic were in some sense or other more *fundamental* than the laws of arithmetic: partly because they dealt immediately with sentences, and also partly because the calculus of logic contained only such words as are used in ordinary language.—You might say, "The same thing applies to mathematics." Yes; but there are terms which we are inclined to call 'mathematical terms'. But "all" and "not" were not originally called 'logical terms'. (Although mathematical terms can be applied in ordinary sentences: "The surface of this elliptical table is integral so-and-so." This is not a mathematical proposition. All the words of mathematics occur *outside* mathematics also. Mathematics gives rules for operating with them. Compare the remark that mathematical propositions are not about numbers.)—So they had the feeling that mathematics *ought* to follow from logic, even before they had any idea how. And in a sense this is entirely correct.

"Russell reduced arithmetical ideas to logical ideas."—In a perfectly good sense arithmetic is logic and also logic is arithmetic.

But consider Russell's idea of *generality*, for example. And also Frege's idea that a numerical statement is about a predicate. This is all right only in a grammatical sense: in so far as in English or German, when we say "There are five circles in this square" or "There have been five thunderstorms in the week", etc., these are predicates only in the sense that we can form a sentence "So-and-so is a circle in the square." In most cases this is not so.

In "There are three circles in the square" we don't say anything about things which are circles and are in the square. *But:* "All the geometrical figures in this square are circles"—then it is the geometrical figures which are circles.

The truth is that the way of writing a generality

$$(\exists x) . \phi x$$

is taken from ordinary language. Only in ordinary language we never say, "There is a *thing* which is a man and has grey trousers." We never talk about bare individuals. We say instead, "There is a *man* who has grey trousers."

It has sometimes been said that the trouble with $(\exists x)$ in ordinary language is that "the number of values is infinite".

For example, "Somewhere in this square there is a circle." You have an idea that this is something like an infinite disjunction of cases.—It has been said that there must be a finite number of places in which a circle can be seen, and it is given as a reason that we cannot distinguish the places which a circle might have below a certain difference (threshold). So when we speak of a circle visible somewhere, this is really a finite disjunction, although a very long one.—But this is bosh. When we say the sentence "Somewhere in this square there is a circle", no disjunction would do for us.

Other people say there is an *infinite* number of disjuncts—thinking that the term has some such meaning as "ad inf." has in mathematics. This is all wrong.—There is no definite number at all. "In how many places can it be?" I don't know. I know I can distinguish masses of different places, and that if I were asked how many I couldn't say. If I talked about "all possible positions of the circle", you would not know what I meant—unless I defined metrically, giving (say) the coordinates of the centre points. But "all possible positions in which one can see a circle" means nothing.

(In certain cases you can give it a sense. Suppose you had a disc which moved slowly across the square, and you asked the subject when he saw a difference of position; and after having moved it up and down, you might multiply the numbers, and say there are a thousand positions in the square. "That's the number of positions which we actually distinguish." Well, if that's what you mean, that's what you mean.)

We use the words "all" and "any" in a mass of different ways of which Russell takes no account.—We can sometimes talk of "all" where we cannot talk of "one". We might say that *all* the points on this panel are painted white, meaning that it is entirely white. But "*one* point is painted white" or "*all but one*"—no one would know what this means at all.

I might have a coloured strip, giving a continuous band of

colours; and I say "All the colours of the rainbow are here." Some would say "An infinite number of colours are here", and some say "finite"; but both would be wrong. "It contains all the shades"—but then the point is just whether you do or do not see finite shades of colour. It is not like 'infinite' in mathematics; it means just that there is not a finite number seen. It is not that "between each shade and another there is always an intermediate shade". We should not know what that meant.—It is *connected* with the mathematical use in the way in which *measuring* and accuracy of measurement is connected with the idea of irrational number.

If Russell gives an interpretation of arithmetic in terms of logic, this removes some misunderstandings and creates others. There are gross misunderstandings about the uses of "all", "any", etc. Russell's propositions are connected with these expressions, but do not do justice to the multiplicity of these uses. There is a confusion between the *uses* of "$(\exists x) . \phi x$" and "$(x) . \phi x$" in ordinary life and their *uses* in mathematics. We have to ask, "What is the criterion of the truth of $(\exists x) . \phi x$, or $(x) . \phi x$?"

What is the criterion
 that all points have been painted white?
 that all men in this room have flannel trousers?
 that all the cardinal numbers are so-and-so?
 that all the colours of the rainbow are from here to here?

These are verified in entirely different ways; their grammar is entirely different. "Surely they have something in common." Not something in common, necessarily; though the uses may have a certain kinship.

Suppose we asked whether Russell has provided *the* calculus for addition of cardinal numbers. For we might say, "Nobody ever said he provided the only calculus. But (a) he gave us an interpretation of our calculus, and (b) he showed what conditions a calculus must fulfil if it is to be a calculus of cardinal numbers. Other sorts of calculus are *short cuts* to this calculus, which is *the* calculus."

If Russell has connected mathematical procedures with logic, this might mean that he just translates them into a new language. But it is misleading to think this an explanation: to think that when we get down to predicates and predicative functions, we see what mathematics is really about.

"We have not got to the bottom of it even yet, but Russell somehow got nearer to the bottom." But what is the bottom? I would say we were at the bottom of it now; a child has got to the bottom of arithmetic in knowing how to apply numbers, and that's all there is to it.

What was it you didn't know before?

If you say that some things have become *clearer* by Russell's work, I agree. But this is a different matter. It means that we see certain connexions more clearly now than before.—It clears up some misunderstandings by illuminating analogies and causes others by misleading analogies. Where an analogy clears things up, this is a great step. If you are able to distinguish things which it has cleared up from things which it has not cleared up—you might almost say you were nearer to the bottom of it. But not automatically.

The analogy of logic being further down is a pernicious analogy.

Why should one want to connect arithmetic with logic? Suppose we said, "Disregard the connexion between arithmetic and logic entirely. Consider arithmetic as a technique which our children learn—perhaps with an abacus." Isn't that all right? Why hanker after logic?

Suppose we wanted to add by Russell's logic instead of the normal way. We say that we shouldn't conclude that the result was correct *just* on the basis of Russell's logic. Then why should a halo attach to Russell's method?

Remember that in deciding whether an addition is correct, we take a certain proposition (of the sort: if one concept is satisfied by . . . , and another concept by . . . —then the disjunction of the concepts is satisfied by . . .)—and decide whether it is a tautology. There are various ways of finding this out. The ordinary way is by cancelling out variables with each other, and we find the whole is a tautology. But if we deal with larger numbers, this will not work any longer. Then it is said, "We can get over this by means of definitions: $1 + 1 = 2$, $(1 + 1) + 1 = 3$, . . ." If we did this, we might get a series of signs such as we have—a numerical *system* like the decimal system. But we might have just —different signs (not signs which form a system). Suppose we use Roman numerals, or some quite different series of signs. Then our additions could no longer be performed as we do them now. We might write a row of numbers up to a certain sign; and then write numbers up to another sign; then correlate the bottom series with the top series, and in this way get the sum. This may or may not lead to the same result as our addition or multiplication. And if the numbers were very big, it is likely it would not.

$$4 \times 10{,}000$$

"If you had a way of transforming, you would be sure of getting 40,000."

What is the sense in this case of saying, "The right result is the one which makes this expression a tautology"? We could in this case give any number of arithmetics, and they would all agree with Russell's definitions and Russell's logic.

Arithmetic is not based on logic. It is based on all sorts of principles which in a sense are logical principles—but not $[p \supset p]$, etc. We can call them laws of thought. Or if you said arithmetic *is* logic—meaning that arithmetical propositions and logical propositions have the same relation to propositions of science—then I'd say yes.

Part of the reason for wanting to say that arithmetic is logic is: we want to get down to propositions.

"Isn't it clear that if I write just '$25 \times 25 = 625$' this does not tell me what it is all about? I see an instrument isolated and disconnected from its use. As though I saw a joiner's plane in mid air. Whereas Russell puts it onto the board."

All I could say is that there is something misleading. Here the instrument has not just one application, as a plane has. You can show, say, one application of it. But in a sense it is *very* much more exact to say "$25 \times 25 = 625$"—this really gives you what you've got. It's just that it *is* in mid air that is good about it.

Take $\sim\sim p \equiv p$
$\sim\sim\sim\sim p \equiv p$

Now imagine we had written down ten thousand $\sim$'s. (You couldn't imagine this. But suppose you saw the whole floor covered; or the road from here to Trumpington.) Now is this one of Russell's formulae? How do you know *which* formula it is? We might say, "If the man counts them correctly—if he does not make a mistake—he will know." But what is a 'mistake' here? We have to introduce something new here in order to know what to call this formula.

"We might try to decide whether it is an odd or an even number." Perhaps—but what are we to take as the criterion that this is an even or an odd number? It does not at all follow from Russell.

We say '$\sim\sim p \equiv p$' is a law of thought. And by analogy to this we may call this long thing on Trumpington Street a law of thought.—But as it stands, it is nothing. We don't know what to do with it.

We need a new rule for reading this sentence. '$\sim\sim p \equiv p$' won't help us. In deciding whether or not this is a formula we make a new calculus.—We can introduce a new principle—say, of counting them. Then we get: that on Trumpington Street there is '$\sim^{1000} p$.' How should we read this? And if I had '$\sim^{1000} \sim^{500} p$', what is this? a negation or an affirmation?

What we are introducing is new methods of calculation, quite analogously to introducing new methods of measurement. We measure the height of a chair. But we measure the height of Mont Blanc in a different way—although there are analogies. But if we measure the distance of the sun from the earth, this is altogether different. We introduce continuously new methods of measurement, and so continuously changing meanings of "length".

This is similar. If we have determined the number of negations, we've introduced a new method. There is (1) a formula which we can overlook (survey); (2) numerals without a system; (3) numerals.

"If we could put rods on end one after another in measuring the distance from the sun . . ."—But "If we could?" What is this? *We don't.* We are constantly using new methods of measurement (new laws of thought). And so it is with Russell.

He wanted to show more clearly in what sense we can say that the laws of arithmetic are laws of thought.—We think in accordance with arithmetic when we say 1500 nuts in one room plus 1600 in the other is 3100 nuts altogether. But what's the use of this? Perhaps I want to know how many nutcrackers I need.

Suppose I wanted to cater for the population of Manchester and Liverpool. That of Manchester is so many hundred thousand, that of Liverpool is . . . So I'll have to cater for two million (say).—I want to calculate how many ovens I need. I will divide the population by so much, and I ought to get the right number of ovens.

Now what does Russell say about this? He says that in order to cater for both Manchester and Liverpool you don't have to put them together but you have to cater for all who are either inhabitants of Liverpool or inhabitants of Manchester. But he can't tell us whether we've added correctly.—The result of the addition cannot be checked simply by saying that in this case we got the right number of ovens. (Whether we got the right number of ovens will depend not on arithmetic simply, but on all sorts of facts: whether anything or anybody vanishes, etc.) But there will be masses of experiences which will check the result. Such-and-such things must come right if we have added correctly. For example, if six intelligent people add, they get the same result. *This* is a point about which Russell says nothing.

If we call a law of arithmetic a law of thought, this means, for example, that if I cater for 30,000 in one town and for 40,000 in another, then I cater for 70,000 altogether.

We are inclined to imagine numbers as *structures.* We could think of a tune as a structure of notes. If we think of a triangle—the number 3 stands for a structure.—We could imagine all of arithmetic done with regular polygons inscribed in circles. "Every number a structure." But as far as *visual* structure goes, there would be no difference between a 1000-sided regular polygon and a 1001-sided regular polygon. If nevertheless we still talk about 'structure', we shall have to change the idea of structure entirely.

"It is some one particular structure, whether we know it or not."
"It has some one number, whether we know it or not."
This is all wrong. What number it has depends entirely on the method of *counting.*
(Think of: "It is either an even number or an odd number, although we may not know it.")

XXIX

I want to say something about logical propositions—considered as laws of thought.

What does a logical proposition *say?* What does it tell us?

You might give different answers.

For example, you write down

$$p \,.\, p \supset q \,.\supset .\, q$$

If you were asked to say what this *says,* you might first read it off, and perhaps translate it into English.—This translation gives us nothing at all.

"It rains."—Suppose someone said, "What does that tell us?"

It would be of no use at all to say, "It says that it's true that it rains." This simply comes to: that it rains.

If a man didn't understand these symbols, you could read them off in English; and this would be an explanation of a sort. Or you could say it means *"il pleut"*, if he can already apply the French sentence but not the English sentence.

But suppose we put real sentences instead of 'p' and 'q'. We may say then that this proposition, being a tautology, says *nothing*. And this is a different kind of explanation.

We might ask, "If you take the propositions of *Principia Mathematica,* what do they say?" Do they say something about English sentences? Or do they say that they themselves are tautologies? Does '$p \supset p$' say that every sentence implies itself, or does it say that "Every sentence implies itself" is a tautology?

The point of Russell's sentences is that none of them gives us any information about anything. If we substitute propositions of botany for 'p' and 'q', then the whole gives us no information about botany; it ceases to be a botanical sentence. This is the point of a tautology: that if any part of it gives information, the rest cancels it out.

Although Russell uses variables: 'p', 'q', etc., he could perfectly well have used ordinary sentences.—Think of demonstrations in Euclid, where nobody thinks we have proved the theorem for *this* circle. In the same way, one *can* perfectly well do algebraic proofs with numbers.—Russell's proofs would lose nothing of their generality, because generality does not lie in what is written down here, but in the way you apply it.

You give a proof here showing that this is a right angle. You apply it to every such angle in a circle.—So we could acknowledge any of Russell's proofs for *any* proposition, although what was written down was some special proposition.

Now we could substitute "It rains" and "I get wet" in '$p . p \supset q . \supset . q$': "If it rains, and it rains implies I get wet, that implies I get wet"—and we call this a *law of thought.*

But isn't this queer?

If I ask you, "What does it say?", you might say, "Something about the weather" or you might say it says nothing. But it doesn't seem to say anything about *thinking.* So why should we call it a law of thought?

Is it a law about the use of sentences? But then it doesn't look like it. Nor does it sound like a rule *about* "and" and "implies". How could it be, when it *uses* these words? and when it talks about raining and getting wet? We might say that the rule is the common structure of all such sentences, but how can such a structure be a rule?

It can be said that these propositions show how we make inferences. We use them in making calculations.

Now does '$p \,.\, p \supset q . \supset . q$' *say* that q can be *inferred* from $p \,.\, p \supset q$? And also does '$p \supset q$' *say* that q can be inferred from p? We are inclined to say this with the former but not with the latter. But the former sentence obviously does not *say* that *q follows* from $p \,.\, p \supset q$, else '$\supset$' would be used differently in the two cases.

We can say then that '$p \,.\, p \supset q . \supset . q$' is not a law of thought. But if you say that this *is a tautology*—then you could call that a law of thought. The law of thought is the statement that that expression *says nothing, all the information is cancelled out,* etc.

And we get the *"follows"* from saying that this is a tautology. We might say, "'$p \,.\, p \supset q . \supset . q$' doesn't say that *$q$ follows* from $p \,.\, p \supset q$—but to say that '$p \,.\, p \supset q . \supset . q$' is a *tautology* is to say that q follows from $p \,.\, p \supset q$."

There is a confusion in Russell, for instance, with '$\sim p . \supset . p \supset q$'. Russell reads this in *words:* "From a false proposition every proposition *follows.*"[1] But does the formula say this? Nothing of the sort.

You can say, "If a proposition is false, then p implies anything", but this does not mean: "From p follows anything."

1. In *Principia Mathematica,* *2.21 reads:
"$\vdash : \sim p . \supset . p \supset q$
I.e. a false proposition implies any proposition." But the number (i.e., chapter) repeatedly translates '$p \supset r$', '$q \supset r$' etc. as "r follows from p", "r follows from q", and so on.

You *can* say, "From $\sim p$ *follows* that p implies q." But what says it follows is the proposition that '$\sim p . \supset . p \supset q$' is a tautology.

One might however say something against this. Take '$p \supset p$ = Taut.' For the moment we are saying that '$p \supset p$' is not the law of thought, but that the law of thought is '$p \supset p$ = Taut.' But Russell never proves this sort of proposition. So wouldn't it be queer to say that Russell after all did not prove laws of thought or laws of logic?

When I wrote, " '$p \supset p$' is a tautology", I made clearer the *application* of '$p \supset p$'.

But why should I not say that Russell, when he affirms '$p \supset p$' means to affirm that it is a tautology?—The question is how he *proves* the proposition '$p \supset p$'. He does not treat "If it rains, it rains" as a proposition about the weather. But if it is proved in a certain way, he treats it as a rule of inference.

We could do logic in a different way. If it is a law of logic that '$p \supset p$' is a tautology, then you might say that it is also a law of logic that '$p \supset q$' is *not* a tautology. So instead of proving that certain propositions are tautologies, you could do logic by proving that certain propositions are *not* tautologies.

But if that is so, and if what stands at the end of the proof in Russell is that proposition—for example, '$\sim p . \supset . p \supset q$'—which we have proved is a tautology—then what would stand at the end of the proof in the other logic, where we prove that certain expressions are not tautologies, would be: '$p \supset q$'. This would mean that at the end of the proof would stand an empirical proposition. It would seem as if I had proved that if it rains, I get wet—and proved it logically.

If Russell is correct in putting at the bottom of his proof '$p \supset p$' instead of '$p \supset p$ = tautology'—then I am correct in putting at the bottom of a proof '$p \supset q$' instead of " '$p \supset q$' is *not* a tautology".

At the end of Russell's proofs we get a tautology. And we do not for a moment doubt how this is to be used. We are inclined to say it is a law of thought.—But if we did logic in the new way suggested, it might seem confusing. On the other hand, it need not be confusing in the least: all that is needed is to know how it is used. So that when we saw at the end of a proof the proposition "If it rains, I get wet", we know that it doesn't say anything about raining and getting wet, but *only* that this is not a tautology.

If you ask, "What does a proposition of *Principia Mathematica say*?"—this is difficult to answer. No one simple answer is correct. If we use sentences in logic, it is not clear offhand what we should say they *say*. It depends (for example) on how long the sentence is. If you read out a longer sentence of *Principia Mathematica*, and if you insert real sentences instead of p and q, and if this is to be a sentence of logic, then its use must be entirely different from the use it seems at first sight to have. It is used in an entirely different way: not to give information, but to show that one can infer something from something else. And therefore Russell could have *said* it in a different way: namely, " '$p \supset p$' is a tautology".—And the example of the new logic I am imagining, makes this still clearer. For here we say experiential propositions—but we do not use them as experiential propositions. A proposition would result, of the form '$p\ .\ q : \supset .r$'; and the whole point of it would be to show that you *can't* infer r from p and q.

That we are to call sentences like Russell's propositions "laws of thought", does not appear on their surface, but it appears by the way they are proved and the way they are used. Actually their form may be said to be misleading. It has actually misled Russell and Frege. Both said they were sentences *about* "logical constants". They said '$p \supset p$' is a sentence *about implication*, as "Lions are fierce" is about lions.

We could even say that the proposition seems to say something which it does not say at all; suggests an application which it does not have. And exactly the same applies to sentences of mathematics and arithmetic.

What does it mean to say that a proposition is a tautology, gives no information, says nothing? We could say various things about this. When I thought about this many years ago, I kept asking, "Why doesn't this say anything?" Then one day I made up the *T-F* symbolism. This symbolism was a means of transforming Russell's propositions into a form where they all looked similar. If you have any of Russell's propositions—either primitive propositions or propositions which follow from these—you always get the same in *T-F* notation: you get a column of *T*'s. I said that this shows it is always true and so cannot be used as an empirical proposition, since it is compatible with every state of affairs.—Can this be called an explanation of "gives no information"? Yes and no. I have translated Russell's propositions into a new symbolism. And we can more easily see from this symbolism that these propositions give no information. But I need not have translated into this symbolism at all. And then I would just have needed to point out the *use* of the propositions.—If you wanted to know the length of a rod, and someone said, "If it is so long, it is so long", would you call this information about the rod? No.

But what does it mean to say it gives no information? We should have to describe all sorts of situations in which one gives or gets information. We should have to describe the language-game: what we do with such a proposition. "If there are fifteen chairs, there are fifteen chairs." In the game in which we ask how many chairs, no behaviour is provided for in answer to this sentence.

I want to order so many chairs for so many people. I ask, "How many people will be present?", I give an order, the chairs are brought, etc. I get information and do a particular thing with it. But if I get the answer, "If fifteen . . . , then fifteen . . .", what could I do with this? Nothing is provided. You might say, "It is just no answer at all; he might as well have said 'Boo boo'." Or you might say, "It is not the same as saying 'Boo boo', since it is an English sentence, and is of the form so-and-so implies so-and-so, and on both sides of the implication sign are real sentences; but although we can distinguish between the two, still the whole thing does not correspond to any actions of mine. There are no actions I take in response to it." If he had said "Fifteen if the weather is fine", there would have been.

Look at the games where information is given. Then you see that this proposition is not used for purposes of information.

Then how is it used?

The sentence 'p implies p' is never used at all. In Russell '$p \supset p$' is used to show certain rules of inference; it itself is not even used as a rule of inference.—You *could* use it thus: You could tell a man that "$\supset$" is a truth function, but not whether it means "or" or "and", etc. Then you tell him that '$p \supset p$' gives no information. And there you have told him something: that "$\supset$" cannot be "or" or "and".

Or you might put it by saying that if you negated '$p \supset p$' you would get a contradiction.

One can express the similarity of laws of arithmetic and of logic—that is, that they are laws of thought—by saying that they impart no information. But one might object: "Logical propositions give logical information; mathematical propositions, mathematical information." (Like: Zoological propositions give zoological information.) We could say this is simply misleading; because if we have a proposition of the form '$p \supset q$', we mean something different by its imparting information than by saying that '$p \supset p$' does.—" 'If it rains, then it's wet' imparts meteorological information. 'If it rains, then it rains' imparts logical information." This is vastly misleading; it misleads us regarding the role which the expression plays.

To say that mathematical propositions impart mathematical information is misleading. For the information they impart is different from what is suggested by their structure as sentences.

What is interesting is not that '$p \supset p$' does not impart information: for any nonsense is similar in that respect—but that certain sentences are put together by certain functions *in such a way* that they do not impart information.

That the lever of the balance doesn't move, gives us information about the weights we have placed in it. But if the lever were fixed, it would give none. Only when we know how "$\supset$" etc. are used otherwise—then it is important that in certain uses they *do not* impart information.

XXX

One might say: Logical propositions are laws of thought in the sense that they might be used to explain the use of certain symbols. Such a law as, for example, '$\sim\sim p = p$' might be used to explain the use of negation. Saying that '$\sim\sim p = p$' is a tautology is a hint as to how the negation sign is to be used. This could be said by saying '$\sim\sim p = p$' follows from the *nature of negation*. But when we say it is in the nature of negation that this gives this—we have the idea of a mechanism behind the symbol.

Every rule you give for the use of a symbol could be given in the form of a mechanism. If I say that the negation sign is to be used in such-and-such a way and not in another, we could give a picture in terms of a mechanism behind what I see, which prevents me from doing this (say a metal $\sim$ moving on rods, etc.)

Any rule can be imagined to be a description of a mechanism —even the rule which says that a pawn must not be moved in a certain way.

"It expresses a law of thought" could be expressed by saying: its being a tautology follows from the way in which we use these signs in thinking, speaking, etc.

Take '$a = a$'. When do we use such expressions? Hardly ever! We could easily do away with all such things. But if we *had* to choose between '$a \neq a$' and '$a = a$', we'd probably all decide for '$a = a$'. Or [we might say,] "If we have a rule '$a = b$', this means that we can substitute a for b in all sentences, and thus in this one."—This shows how '$a = a$', although utterly useless, is the natural thing to say.

I can say of the colour of one book that it is the same as the colour of another book. But it doesn't follow that we can say "This book = this book."—[We might imagine] two books which coincide more and more: you could say they become more and more the same—until when they coincide you could say, "This is the same as this."

If we do use "identity" as we do, then it's natural that we extend our use up to this point, saying "This is this", although this is perfectly stupid and useless.

Take the case of a white chalk mark on a blackboard. Should we say that it *fits* into its surroundings? It may be said that it is a *picture* of something fitting into something. With our actual use of "fit" it doesn't make sense to say that the white dot fits into the black surroundings. But if we had to say one or the other we'd all be inclined to say it does fit rather than that it doesn't. This would be a more natural extension of our use of "fits".

"Every coloured patch fits into its surroundings." This is exactly the same as the law of identity. Or: "Everything fits into its own region of space."—If someone were to say this, we know what kind of picture he has. And we might allow him to say this; we might say it is a law of thought and expresses something about the use of the word "fits".

(1) '$a = a$' is a perfectly useless proposition.

(2) It is suggested by the other uses of "equals", etc.

The idea of a law of thought is similar to the idea of a law of measurement. If I describe the rules of measurement you might regard these rules as rules given to someone, so he knows what to do—but all kinds of important things can be inferred from them. That a grocer weighs cheese [in such-and-such a way] is interesting, because it shows that the weights of things don't constantly fluctuate.

You could call the *rules of weighing* laws of thought. In a way, they *define "weight"*.

'$p . p \supset q . \supset . q$' could be called a law of thought because it allows us to infer this kind of proposition from this kind. The whole point is that no experience should be allowed to make this inference valid or invalid.

$$(n)^{\phi}$$

This is short for: "There are n things satisfying this condition (satisfying the condition ϕ)."

$$(n)^{\phi} \,.\, (m)^{\psi} \,.\, \text{Ind.}$$

means that there are n things having ϕ and m things having ψ and nothing which is ϕ is ψ.

$$\overset{4}{(n)^{\phi}} \,.\, \overset{5}{(m)^{\psi}} \,.\, \text{Ind.} \supset .\, (n \overset{9}{+} m)^{\phi \vee \psi}$$

Now this is a tautology. It's often been thought that this is what is meant by "4 + 5 = 9". But this is wrong. Because if we write

$$\overset{4}{(n)^{\phi}} \,.\, \overset{3}{(m)^{\psi}} \,.\, \text{Ind.} \supset .\, (n \overset{9}{+} m)^{\phi \vee \psi}$$

this ought to be a contradiction; which it isn't. Suppose that the antecedent were true. It would simply be a case of a true proposition implying a false and would be simply false.

Take '$(4)^{\phi} \supset (5)^{\phi}$'. Is this a contradiction? Of course not. "(Four people are in this room) $\supset$ (five people are)" says "$\sim$ (Four people are in this room) **v** (five people are)". '$p \supset \sim p$' is not contradictory. It simply says $\sim p \vee \sim p$, which equals $\sim p$.

But the proof that

$$\overset{4}{(n)^{\phi}} \,.\, \overset{5}{(m)^{\psi}} \,.\, \text{Ind.} \supset .\, (n \overset{9}{+} m)^{\phi \vee \psi}$$

is a tautology, as apart from the assertion of it, is equivalent to the mathematical proposition. [1]

Suppose we take

$$(n^{\text{3 billion}})^{\phi} \,.\, (m^{\text{4 billion}})^{\psi} \,.\, \text{Ind.} \supset .\, (n + m^{\text{7 billion}})^{\phi \vee \psi}$$

1. This sentence is based on S. The corresponding sentence in M reads: "But the proof that the above proposition [M is referring to

$$\text{'}\overset{4}{(n)^{\phi}} \,.\, \overset{3}{(m)^{\psi}} \,.\, \text{Ind.} \supset .\, (n \overset{9}{+} m)^{\phi \vee \psi}\text{'}]$$

is not a tautology is a proof that 4 + 3 doesn't equal 9."

Is it clear that this is a tautology? Now in Russell's symbolism in proving small calculations we have '$(x)(y) \supset (xy)$'. From this we can easily make a tautological form. But if we have a large number of variables, we don't know how to do this.

If we say there are two billion in Europe and three billion in America, are we entitled to say that there are five billion altogether? Now we can say that in saying this I made a correct *inference.* But although we all agree in this, we don't try to see whether it's a tautology.—We could *extend* the use of "tautology". *Now* we don't say that if so-and-so is a tautology an inference can be made, but that since an inference can be made in such-and-such a way, this is a tautology.

When we wrote out all the signs for variables, we might make new signs: we might put "Jack" in occasionally. Or we might use the cardinal numbers instead of letters, change the order, etc. —Suppose we have a bracket with three billion names, one with four billion in it, and another with seven billion. Now we have to devise a calculus, a method of calculation, for cancelling out these enormous rows of names.

Russell doesn't provide us with a technique for doing this. Now is there only one technique? How do we decide between two techniques which conflict? What reason have we to choose one rather than the other?

We have a proposition of Russell's, and it should follow from Russell's axioms. This is what we should be guided by. But how do we know when it follows from Russell's axioms? Suppose Malcolm and Wittgenstein get different results with their different techniques. Why should we prefer one to the other? A new technique is introduced, and we have no reason for preferring one to the other. We have to make a decision as to which technique to use.

We determine whether this huge expression is a tautology by getting the sum of the numbers of names on the left and comparing with the number of names on the right. But how do we get the sum? What makes us decide which technique for getting the sum is to be used?

It is not like the case where we have a mechanism with wires, where a bell rings when we get a tautology.[2]—We wouldn't for a moment think that the result was 2 billion + 3 billion = 5 billion, *because* we have made this calculation. We choose a calculus not at all because it agrees with Russell, but rather we'd choose Russell at any time because it agrees with the decimal way of multiplying or adding.

We might check up on 3 + 4 = 7 this way:

1——1
2——2
3——3

1——4
2——5
3——6
4——7

We draw lines and make a one-to-one correlation. Now it is clear that we won't do the same with 3 billion + 4 billion = 7 billion.

If in one case (a large number) we added and got a certain result, and then correlated and got a different result, we should certainly trust the addition rather than the correlating. We should say that there *must* be a mistake in making the dashes. Hence, not only have we never used the dashes to establish such a result, but it is a fact that we should always trust the calculation rather than the system of dashes.

This is all I'm saying. We already have a calculus and we don't check up on it by some other method. Instead, if anything disagreed with this calculation we should reject it.

We have two different methods of adding here; someone might use either method in practical affairs.—This really means we have two different meanings of "sum"—because in this case we take *this* as infallible, our criterion; in the other case, *this*.

Thus in Russell's case: if we use some method of proof other than ordinary calculation (addition) to show that 31478 + 97 =

2. See Lecture XXVII.

31575, and these two disagree, we should almost certainly take the method of ordinary addition. But this means that there is really no question of checking up on ordinary arithmetic.

Russell's system is built on the fact that with small numbers there is almost always agreement between different methods of getting the number. But if we came to consider cases where there might be disagreement in the methods or *standards,* we see that we wouldn't allow a new standard to replace or refute the results of our ordinary standard of *addition.*

A calculation is like a standard of measuring. If we used dashes and correlated with them and this didn't agree with the calculation, we might say that we had made a mistake, or one had vanished or something. If we had to choose we would choose the calculation, and we would even check up on the other by *it.*

And now Russell's logic has nothing to do with our making this choice. Our ordinary calculus is entirely independent of Russell's logic.

XXXI

Let's consider a proof by mathematical induction.

ϕ is a mathematical property. If it is known that $\phi(1)$, and it is known that $(n): \phi(n). \supset .\phi(n+1)$, then $(n) . \phi(n)$. It's misleading because it's not known how $(n): \phi(n). \supset .\phi(n+1)$ is proved.

Suppose we know:

$$\phi(1) = \psi(1)$$
$$\phi(n+1) = F(\phi n)$$
$$\psi(n'+1) = F(\psi n')$$

We can then substitute 1 for n, and so prove that $\phi(1+1) = \psi(1+1)$, and then substitute $1+1$ for n and so prove that $\phi(3) = \psi(3)$, . . .

On the ground of these three equations, we can assert $\phi(3000) = \psi(3000)$. We could prove this by going through all 3000 steps, but this would be very long. So we take a short cut.—Now what's responsible for this? How is it that we can leave all these steps out? Someone might say, "This is rash." But of course it isn't. We can even conclude to $\phi(n) = \psi(n)$, which if anything would be more rash.

The interesting thing is this: (1) You haven't the faintest doubt you can do this. (2) You can say you have done the same thing in the two cases. Your mind has gone through all the steps.—But has your mind done this?

What did I do when I tried to persuade you of this? I showed you what would happen with 1, what would happen with 2, and then said "and so on". And you were satisfied with this.

You might say:

(1) The real proof would be the whole chain. In some mystical way, I've done all these operations.

(2) In some way what I've done is the *same as* doing all 3000.—I can only say it's not the same.

Now how is it that we can be so certain that the two steps will lead to the same result as the 3000 steps? Why do we say they must meet?

Suppose we have a line cut by an Archimedean spiral at equidistant points.

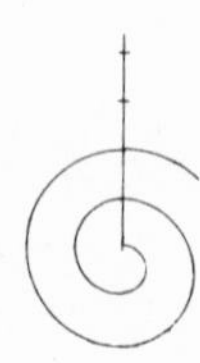

We measure to three of these points and then measure the line and find that it consists of a hundred such segments. We then say that if we make a hundred spirals we'll reach the end of the line. How are we so certain of this? Doesn't it seem queer that we can neglect all the steps and suddenly be there?

We say, "If the spirals are drawn correctly it *must* come to that point."—This is a declaration of geometry—[and tells us nothing about the world]. It is a new rule. If we continued the spirals with a gauge and it didn't come right, we might say that the spiral wasn't Archimedean or that something had happened to our gauge.[1]

Are we taking a criterion for each point? Or do we take this thing at the end as a criterion? Are the measures independent or not?

If we weigh two crystals separately and each weighs 3 grams, and then weigh them together, must they weigh 6 grams? Obviously not. They might weigh 5 grams. Does this mean that our arithmetic has gone wrong? Of course not.—Our arithmetic doesn't tell us anything about the weights—it is a standard. On the basis of it we judge that something *must* have happened.

So in the case of mathematical induction. How are we going to judge that we made the right steps? If we should go through the 3000 steps and *not* get the predicted result, we're going to say that we've made a mistake. Either we take this as one of the criteria or we don't.

Isn't it all that you can say of a step that it seems to be correct? What is it to say of a step that it is correct, except to say "It seems to be correct—unless I am crazy"?

Take: 3 angles of a triangle = 2 right angles. We make measurements, between stars, for example, and we don't get two right angles.

(1) "It was a mistake of measurement."

(2) Somebody might say, "Our measurement *was* correct—but we haven't measured along straight lines. The whole thing is comparable to measuring on a sphere."

In (1) we say, "I'm going to take coming to 180° as one of the criteria of measurement." For some reason we decide to use a new criterion for correctness of our measurement.

In the mathematical induction case, you might say, "We *must* get the same result."

1. (From "How are".) Neither version of this passage is complete; the order of the material taken from both has been altered.

Does this mean that we *do* get the same result? The thing is that we get this result, and *not* by taking the 3000 steps; and in fact we use this result as a criterion of whether our 3000 steps *are* correct.

That our 3000 steps will produce this result is an experiential proposition. But that this result is correct is a *rule*. We don't allow any experiential process either to refute or to establish a rule.

What has this to do with the point about adding according to Russell and adding according to the decimal system?

Now an addition according to the decimal system can easily be proved. This might be done step by step (1 + 1 + 1 + 1 . . . etc.) or by induction. And what's been said above is to be noted here.

Suppose we have a formula of Russell's, ( )( ) ⊃ ( ) with an enormous lot of variables in the brackets, and we prove in Russell's way that this is a tautology, that it follows from the primitive propositions. We then count the variables and we get [the sum by ordinary addition]—and what we get [with Russell] doesn't agree with what we get in the decimal calculus. It may or may not be that if this comes out to be a tautology, this number plus this number gives that.

If it doesn't come out and we say that we've made a mistake—then we've applied the decimal calculus as a new criterion for the validity of the proof; and this is a criterion which Russell hasn't thought about.

We should in the great majority of cases get the same result. —If we're tired, we leave out numbers. But I've assumed a criterion for leaving out numbers. If we all were to do this, there'd be no one to notice it.—If we had to count very large numbers of signs, we should very probably always get different results. We might say, "In some way not known to us, we've left out numbers"—where the criterion is different results.

We could quite easily get to the point where we would trust a multiplication to tell us we'd made a mistake in counting which we couldn't possibly find. But would we ever allow anything to show that we'd made a mistake in multiplying that we couldn't possibly find? We shouldn't ever allow anything to prove that we're wrong in saying $12 \times 12 = 144$. For this is what we *call* correct multiplication. Nothing could prove it *incorrect*, in the way in which a multiplication can prove a counting incorrect.

Suppose someone says, "Surely we have a mathematical proof that they *must* come to the same." Or "If the multiplication is correct and if the counting is correct, then there can't be a disagreement. We can't imagine there to be a disagreement."—The difficulty is to make a cut in such a way as to cut the mathematical from the empirical.

It isn't *a priori* clear at all when we should say we've left out numbers. A criterion is that we say we do the same thing.—Our memory agrees with what we write.

We don't know what would happen if things went wrong, or what sort of mistake we should imagine.

There is a case where we all say you left out a number and you don't know you have. What we say is the criterion.

"Don't you see you've left out the number 1000?" You look at the text book. You say, "I must have been dreaming."—Did you ever hear of such a case? Did you ever hear of a man imagining that the number series goes the other way? No. But these are very important facts. If *many* people did such things, this would affect the nature of our calculus. The criterion for our counting correctly is partly our memory, but mostly the constant agreement.

If we say, "What are the facts that join up Russell's calculus with the decimal calculus?" or "What are the facts on which our calculus is based?"—I would say: They are the empirical facts that we generally remember mistakes in counting, don't twist the calculus around, and so on. These are extremely general facts, hardly noticed.

Now is it an empirical fact that $\phi(3000) = \psi(3000)$?—We *make* the rule that $\phi(3000) = \psi(3000)$ *because* of the agreement in action—namely, that if we went through these steps we would nearly all get the same results. And this rule then becomes a standard of measurement. The rule doesn't express an empirical connexion but we make it because there is an empirical connexion.

We get the same result as in the mathematics books. If we don't get the same, we either (1) find a mistake, or (2) if we don't find the mistake, we say that because of the disagreement, there *must* be a mistake.—This whole thing is an empirical fact. Now what is the mathematical fact?

Two propositions: what is the difference?

(1) If we weigh one body in such-and-such a way, and another body in such-and-such a way—then if we put them both on a balance in such-and-such a way, we always get the sum of their weights.

(2) If we weigh one body in such-and-such a way, and then another—then if we put them both on a balance in such-and-such a way, and don't get the sum of their weights, we *must* have made a mistake.

The "*must*" when it appears like this always indicates a rule. The first is an empirical proposition, the second a mathematical one.

When do we say we've weighed correctly?—We have introduced a new criterion.

We are so used to the criteria for certain facts, that we completely forget what the criteria are. We need an enormous number of criteria for knowing that we count the same, etc.

There is one more point, enormously important, which I won't make clear.

At first sight it sounds unthinkable that counting in one way should lead *there* and counting in another way *there.* "Surely something must have been incorrect."

With respect to calculations like multiplication, addition and so on: we are all inclined, if calculations like these are meant, to take particular kinds of examples as paradigms.

One reason for the difficulty in imagining that we should not know whether something has gone wrong with counting or not is that we take as our paradigm of calculation $2 + 2 = 4$. We think of cases where we can see at a glance, where there is the criterion of a visual group, which doesn't exist later on.

What would it be like if you calculated wrongly? (Even here you might get into difficulties.)

We take calculations with fingers, or with numbers which have a particular face, which the number 10,000 doesn't have.

You can give the number 10,000 a face, arrange points, etc. You might have a case where a calculation made by digits and calculation made by a face wouldn't agree.

The seed I'm most likely to sow is a certain jargon.

Index